operational amplifiers
and
linear integrated circuits

second edition

operational amplifiers
and
linear integrated circuits

Robert F. Coughlin

Frederick F. Driscoll

Wentworth Institute of Technology

PRENTICE-HALL, INC. *Englewood Cliffs, New Jersey 07632*

Library of Congress Cataloging in Publication Data

Coughlin, Robert F.
  Operational amplifiers and linear integrated
  circuits.

  Bibliography: p.
  Includes index.
  1. Operational amplifiers. 2. Linear integrated
circuits. I. Driscoll, Frederick F.
II. Title.
TK7871.58.06C68   1982   621.3815′35   81–4334
ISBN 0–13–637785–8   AACR2

Editorial/production supervision
  and interior design by Lori Opre
Cover design by Jorge Hernandez
Manufacturing buyer: Gordon Osbourne

© 1982 by Prentice-Hall, Inc.
Englewood Cliffs, N.J. 07632

Printed in the United States of America

10 9 8 7 6 5 4

PRENTICE-HALL INTERNATIONAL, INC., *London*
PRENTICE-HALL OF AUSTRALIA PTY. LIMITED, *Sydney*
PRENTICE-HALL OF CANADA, LTD., *Toronto*
PRENTICE-HALL OF INDIA PRIVATE LIMITED, *New Delhi*
PRENTICE-HALL OF JAPAN, INC., *Tokyo*
PRENTICE-HALL OF SOUTHEAST ASIA PTE. LTD., *Singapore*
WHITEHALL BOOKS LIMITED, *Wellington, New Zealand*

To Bae *and* Fred
and to
Millie *and* Wally

contents

2 FIRST EXPERIENCE WITH AN OP AMP   *10*

3 INVERTING AND NONINVERTING AMPLIFIERS   *31*

4 COMPARATORS   *51*

5 SELECTED APPLICATIONS OF OP AMPS   74

7 OP AMPS WITH DIODES   *131*

8 DIFFERENTIAL, INSTRUMENTATION, AND BRIDGE AMPLIFIERS   *154*

9   BIAS, OFFSETS, AND DRIFT   *181*

10 BANDWIDTH, SLEW RATE, NOISE, AND FREQUENCY COMPENSATION   *200*

11 MODULATING, DEMODULATING, AND FREQUENCY CHANGING WITH THE MULTIPLIER *217*

12 ACTIVE FILTERS   *247*

13 INTEGRATED-CIRCUIT TIMERS   *273*

14 POWER SUPPLIES AND POWER AMPLIFIERS *304*

preface

Operational amplifiers and other linear integrated circuits are both fun and easy to use, especially if the application does not require the devices to operate near their design limits. It is the purpose of this book to show just how easy they are to use in a variety of applications involving instrumentations, signal generation, filters, and control.

When first learning how to use an op amp, one should not be presented with a myriad of op amps and asked to make an informed selection. For example, one manufacturer presently lists 130 different op amps.

Our introduction begins with an inexpensive, reliable op amp that forgives most mistakes in wiring, ignores long lead capacitance, and does not burn out easily. Such an op amp is the 741, whose characteristics are documented in Appendix 1 and whose applications are sprinkled throughout the text.

If a slightly faster op amp is needed for a wider bandwidth, another inexpensive and widely used op amp is the 301. See Appendix 2 for its electrical characteristics and Chapter 10 to learn when one might prefer the 301 over the 741.

Chapters 1 to 8 show how the op amp can be used in a number of applications. The limitations of op amps are not discussed in Chapters 1 to 8 because it is very important to gain confidence in using op amps before pushing performance to its limits. When studying transistors or other devices, we do not begin with their limitations. Regretably, much of the literature on integrated circuits begins with their limitations and thus obscures the inherent simplicity and overwhelming advantages of basic integrated circuits over basic transistor circuits. For these reasons, op amp limitations are not presented until Chapters

9 and 10. Furthermore, not all op amp limitations apply to every op amp circuit. For example, dc op amp limitations such as offset voltage are usually not important if the op amp is used in an ac amplifier circuit. Thus dc limitations (Chapter 9) are treated separately from ac limitations (Chapter 10).

A fascinating integrated circuit, the multiplier, is presented in Chapter 11, because it makes analysis and design of communication circuits very easy. Modulators, demodulators, frequency shifters, a universal AM radio receiver, and a host of other applications are performed by the multiplier, an op amp, and a few resistors.

In Chapter 12, the four basic types of active filters are shown: low-pass, high-pass, band-pass, and band-reject filters. Butterworth filters were selected because they are simple to design. If you want to design a three-pole (60-dB/ decade) Butterworth low- or high-pass filter, Chapter 12 tells you how to do it in four steps with a pencil and paper. No calculator or computer program is required. Basic algebra is the only mathematics that is required throughout the text.

Chapter 13 is devoted to another class of ICs that is widely used and as versatile as the op amp—IC timers. They are inexpensive and simple to use.

Chapter 14 provides an introduction to IC voltage regulators. It begins by showing how the op amp and a few components not only can regulate voltage but can also protect the regulator circuit from overloads. This shows the reader the natural evolution of op amp regulator circuitry into the modern, inexpensive, simple-to-use IC voltage regulator. Of all the available IC regulators (one manufacturer lists 72 types), we have chosen three types: (1) an adjustable positive-voltage regulator, the LM317; (2) an adjustable negative-voltage regulator, the LM337; and (3) a dual-tracking regulator, the RC4195. These three regulators will solve most of the normal requirements for low-power regulators that you will encounter.

This second edition has been revised to include more examples of how linear ICs are used. Chapter 7 is entirely new. It was needed to show how op-amp and diode combinations could sample, precisely rectify, separate polarity, and otherwise condition signals for further processing. Other sections were inserted to give more detail on instrumentation or the need for specific circuits. For example, Chapter 8 contains instruction on the techniques for successfully measuring strain with strain gages. Chapter 5 reviews the disk-record recording process to show why you need a phonograph preamplifier to enjoy a record.

This text has enough material for a one-semester course. All the circuits have been classroom and laboratory tested and have been selected with great care to illustrate (only partially) the tremendous range of applications for which ICs provide inexpensive solutions to practical problems. The material is suitable both for nonelectronics specialists who just want to learn something about linear ICs and for electronics majors who wish to use linear ICs.

We thank Mrs. Priscilla Booth and Mrs. Phyllis Wolff for their assistance in the preparation of the revised manuscript. Our colleagues Robert S. Villanucci, William Megow, John Marchand, John Cremin, Richard Bean, Dominic Giampietro, and Alexander Avtgis have been particularly helpful with their constructive criticism, helpful suggestions, and testing of ideas.

Finally, we thank our students for their insistence on relevant instruction that is immediately useful.

ROBERT F. COUGHLIN
FREDRICK F. DRISCOLL

Boston, Mass.

breadboarding
op amp circuits

ᵣᴎᵢᵣⁱ

1

1-0 INTRODUCTION

Linear integrated circuits (ICs) are being incorporated almost daily into more and more electronic applications in such fields as audio and radio communications, medical technology, instrumentation, manufacturing controls, and automotive technology. The reasons for their growing use are small size, ease of use, reliability, and low cost. Circuits that just a few years ago required weeks of design time can now be purchased as ICs. This allows the ICs to be used as a functional block. With a few external components added to the IC, the design is completed. This functional block is also much simpler to analyze and troubleshoot than its discrete-component equivalent.

There are two types of linear integrated circuits manufactured today, namely *monolithic integrated circuits* and *hybrid integrated circuits*. Monolithic ICs are manufactured on a single semiconductor substrate (usually silicon) along with some resistors and capacitors. Hybrid ICs are a combination of discrete components, passive elements, and monolithic ICs. All circuits in this text use a monolithic IC.

This book is an introductory text, but all the circuits have been laboratory tested. They have been chosen to demonstrate a wide variety of applications and to illustrate the basic characteristics being discussed at the time. Many readers of a text wish to set up the circuits to check results and formulas and to prepare for study in more depth. Section 1-1 shows two types of homemade linear IC breadboards that allow linear IC circuits to be set up quickly and easily. Many other types of breadboards are commercially available.

1-1   LINEAR INTEGRATED-CIRCUIT BREADBOARDS

1-1.1   Breadboard Requirements

A breadboarding system should allow you to:

1. Set up a circuit in a minimum amount of time (approximately 2 minutes for most of the circuits in this text)
2. Change external elements easily for any design modifications
3. Replace the IC without having to unsolder it
4. Connect leads from audio oscillators, function generators, voltmeters, and oscilloscopes
5. Have a positive and negative power supply

One type of homemade breadboard is described that uses two 9-V batteries for the power supply. A second type uses a plus/minus regulated power supply, the design of which is included in Chapter 14.

1-1.2   Battery-Operated IC Breadboard

Figure 1-1 shows an inexpensive IC breadboard that allows experimentation with a 741 operational amplifier. The op amp is soldered to the bottom side of

Figure 1-1   Inexpensive breadboard for a 741 op amp.

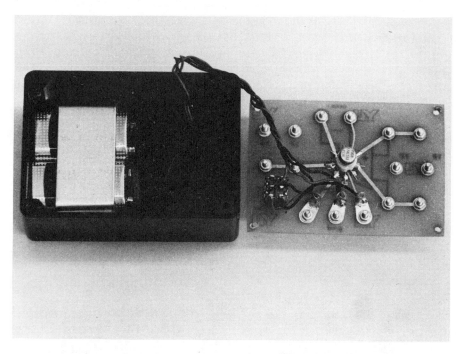

Figure 1-2 Bottom view of IC breadboard.

Figure 1-3 Linear IC breadboard.

the printed-circuit board as shown in Fig. 1-2. When the power switch is thrown to "on," respective voltages of +9 V and −9 V are applied to pins 7 and 4 from the two 9-V batteries. When the switch is in the "off" position, power can be supplied to the op amp terminal springs from an external supply. Advantages of this breadboard are portability, low cost, terminals that are easily available, and extra tie points. Its limitations are that it has been designed for only a particular IC (the 741 or 301 op amp) and that the batteries will eventually have to be replaced.

1-1.3   IC Breadboard with Regulated Power Supply

A more versatile linear IC breadboard is shown in Figs. 1-3 and 1-4. It consists simply of a box frame, two printed-circuit (pc) breadboards, and a

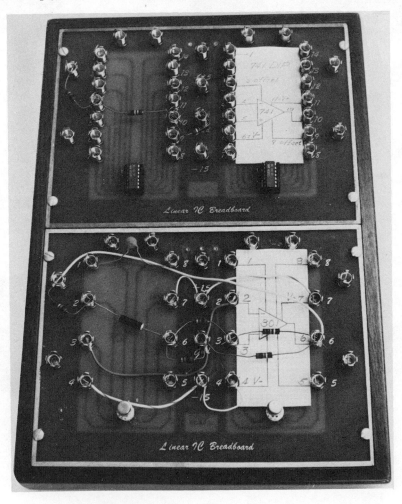

Figure 1-4   Using the linear IC breadboard.

±15-V power supply. The power supply is mounted on the inside rear wall as shown in Fig. 1-5. The undersides of the breadboards are shown in Fig. 1-6.

The three center terminals of the breadboards provide +15 V, ground, and −15 V. Figure 1-4 shows the breadboarding arrangement for three of the circuits studied in the text. The bottom two ICs are wired for the triangular-wave generator in Fig. 6-14. The upper left IC is the inverting amplifier of Fig. 3-1.

As shown in Fig. 1-4, a paper template can be made for the op amp pin connections to identify input, output, power supply, and other terminals. A template for a 741 dual-in-line package op amp is shown for the top right IC in Fig. 1-4; the bottom right template is for a 301. For more experienced users, the template can be omitted. Further details on this linear IC breadboard are provided in Section 1-2.

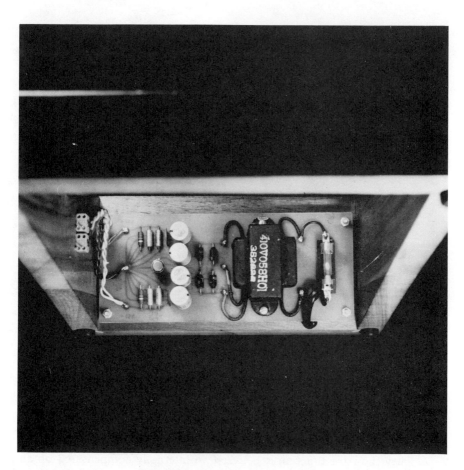

Figure 1-5 Power supply mounting.

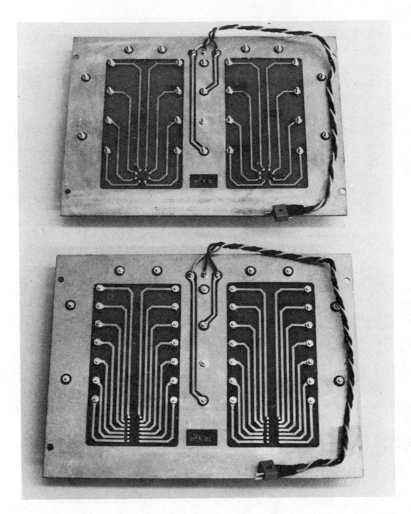

Figure 1-6 Undersides of breadboards.

1-2 HARDWARE FOR THE LINEAR BREADBOARD

1-2.1 ICs and IC Sockets

IC sockets can be purchased for all the standard ICs. The op amps chosen for most of the circuits in this text are the 741 and 301 (see Appendices 1 and 2). The reasons for choosing these two devices are:

1. They are general-purpose op amps.
2. They show the differences between an internally compensated op amp, the 741, and an externally compensated op amp, the 301 (see Chapters 9 and 10).

3. They are inexpensive, costing less than $1.
4. Both are readily available in a 14-pin dual-in-line package (DIP) and an 8-pin round case (TO99). See Appendices 1 to 3 for details on pin counting and package styles.

1-2.2 Spring-Clip Connectors

The spring clips (Omni-Grip; U.S. Patent No. 2,150,911; courtesy of Cosmic Voice, Inc.) shown in Figs. 1-1 to 1-4 were chosen for the following reasons.

1. They accept most wire sizes used in electronic work.
2. Components with different lead thicknesses can be connected to the same spring clip.
3. They are large enough to work with easily.
4. They will accept alligator and spade connectors.
5. One spring clip can be chosen as the common or ground terminal, thus eliminating ground loops.

1-2.3 Breadboard Material List

The materials used for the linear IC breadboard in Figs. 1-3 to 1-5 are:

1. *Box frame* or chassis, as available; these were made by stripping a vacuum-tube power supply and sawing the box in half at an angle
2. *Omni-Grip spring clips*, 40 for DIP board and 28 for 8-pin metal can board (available from General Electronics Associates, Inc., P.O. Box 156, Northfield, Ohio 44067)
3. *IC sockets*, four 14-pin for DIP board or four 8-pin for metal can
4. *Mounting hardware* 4/40 nuts and bolts
5. G10 double-sided printed-circuit board
6. *Power supply*
   Option 1: two 9-V batteries
   Option 2: buy a ±15-V 50-mA power supply for approximately $40
   Option 3: build the supply of Fig. 1-5 from information given in Figs. 14-2 and 14-6 with the following equipment (approximate cost: $15)
   One Raytheon RC4195T fixed ±15-V dual-tracking voltage regulator
   One 110-V to 30-VCT at 1 A transformer
   Four 1-A diodes
   Four 500-μF electrolytic capacitors at 50 WV dc for unregulated supply
   Four 5-μF electrolytic capacitors, 35 WV dc tantalum, for output of the 4195
   Two 10-kΩ, $\frac{1}{4}$-W bleeder resistors, one from $+V_o$ to ground and one from $-V_o$ to ground of the 4195
   One fuse holder and 0.5-A fuse
   One line cord plus hookup wire

1-3 USING THE IC BREADBOARD

1-3.1 Breadboarding an IC Circuit

After you have designed and analyzed a circuit containing an IC and obtained all the components, you are ready to build and test your design. With an IC breadboard similar to those described in Figs. 1-1 to 1-6, the time needed to build a prototype should be only a matter of minutes.

Insert the IC and, with power off, connect the $+15$-V and -15-V terminals on the breadboard to the $+V$ and $-V$ terminals of the IC. Connect all grounded leads to the single ground spring clip, if possible. Complete the circuit by connecting all other components to the proper spring clips. Use tie points if necessary. Now, and most important, *recheck* your entire layout; if you are like the rest of us, you have already made a mistake. Try to keep leads short.

1-3.2 Testing an IC Circuit

After you have checked your wiring layout, you are ready to proceed.

1. Connect the power by plugging in the line cord (not shown).
2. With a CRO on the 10-V/cm scale and dc coupled, or with a dc voltmeter, check to be sure that the $+V$ terminal is at $+V$ and the $-V$ terminal at $-V$.
3. Connect the input signal and measure it.
4. Measure the output voltage to be sure the output is neither in saturation (see Chapter 2) nor slew-rate limited (see Chapter 10). If the input signal is ac, the output must be checked on an oscilloscope to be sure that the output is not overdriven or distorted.
5. Take all measurements with respect to ground. For example, if a resistor is connected between two terminals of an IC, do not connect either a meter or a CRO across the resistor; instead, measure the voltage on one side of the resistor and then on the other side and calculate the voltage across the resistor.
6. Unless otherwise instructed, avoid using ammeters. Measure the voltage as in step 5 and calculate current.
7. Load resistors should not be less than $2\,k\Omega$; the reason for this is discussed in Section 2-1.2.
8. Disconnect the input signal before the dc power is removed. Otherwise, the IC may be destroyed.
9. These ICs will stand much abuse. But *never:*
   a. Reverse the polarity of the power supplies,
   b. Drive the op amp's input pins above or below the potentials at the $+V$ and $-V$ terminal, or
   c. Leave an input signal connected with no power on the IC.
10. If unwanted oscillations appear at the output and the circuit connections check out,

a. Connect a 0.1-μF capacitor between the op amp's $+V$ pin and ground and another 0.1-μF capacitor between the op amp's $-V$ pin and ground,

b. Shorten your leads, and

c. Check that test instrument, signal generator, load, and power supply ground leads come together in one spring clip (star grounding).

11. The same principles apply to all other linear ICs.

We now proceed to our first experience with an op amp.

first experience
with an op amp

ᴘᴜ

2

2-0 INTRODUCTION

The name *operational amplifier* was given to early high-gain amplifiers designed to perform the mathematical operations of addition, subtraction, multiplication, and division. They worked with high voltages such as ± 300 V, but they could accomplish computations, such as solving calculus problems, that were not economical to undertake prior to their invention.

The modern successor of those amplifiers is the *linear integrated-circuit op amp*. It inherits the name, works at lower voltages, and is at least as good. Today's op amp is so low in cost that millions are now used annually. Their low cost, versatility, and dependability have expanded their use far beyond applications envisioned by early designers. Some present-day uses for op amps are in the fields of process control, communications, computers, power and signal sources, displays, and testing or measuring systems. The op amp is still basically a very good high-gain dc amplifier.

One's first experience with a linear IC op amp should concentrate on its most important and fundamental properties. Accordingly, our objectives in this chapter will be to identify each terminal of the op amp and to learn its purpose, some of its electrical limitations, and how to apply it usefully.

Op amps have five basic terminals: two for supply power, two for input signals, and one for output. Internally they are complex, as shown by the schematic diagram in Fig. 2-1 and Appendix 1. It is not necessary to know anything about the internal operation of the op amp in order to use it. The people who design and build op amps have done such an outstanding job that external components connected to the op amp, not the op amp itself, determine what it will do.

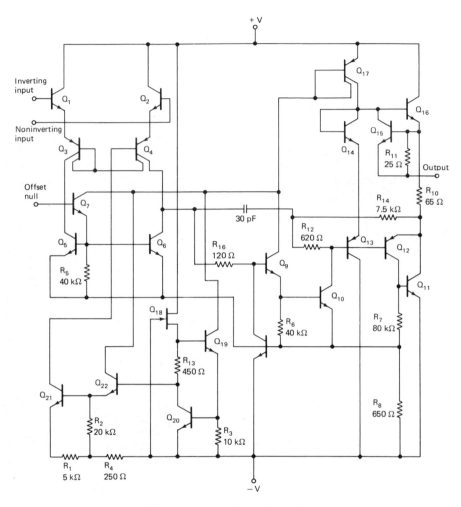

Figure 2-1 Op amp schematic.

Four common packages are shown in Fig. 2-2 and in Appendix 1. As viewed from the top, pins are counted in a counterclockwise direction, similar to vacuum tubes. Pin 1 is identified by a notch on the DIP of Fig. 2-2(c) and (d) and by a dot on the flat pack of Fig. 2-2(b). Pin 8 is identified by a metal tab on the metal can package of Fig. 2-2(a).

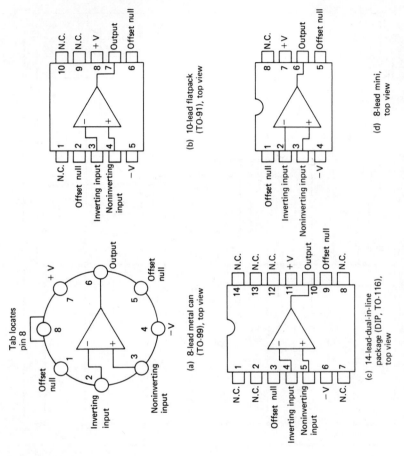

Figure 2-2 Connection diagrams for typical op amp packages. The abbreviation N.C. stands for "no connection." That is, these pins have no internal connection, and the op amp's terminals can be used for spare junction terminals.

(a) 8-lead metal can (TO-99), top view

(b) 10-lead flatpack (TO-91), top view

(c) 14-lead-dual-in-line package (DIP, TO-116), top view

(d) 8-lead mini, top view

2-1 OP AMP TERMINALS

The circuit schematic for the op amp is an arrowhead, as shown in Fig. 2-2. The arrowhead symbolizes amplification and points from input to output.

2-1.1 Power Supply Terminals

Op amp terminals labeled $+V$ and $-V$ identify those op amp terminals that must be connected to the power supply; see Fig. 2-3 and Appendices 1 and 2. Note that the power supply has *three* terminals: positive, negative, and power supply common. The power supply common terminal may or may not be wired to earth ground via the third wire of line cord. However, it has become

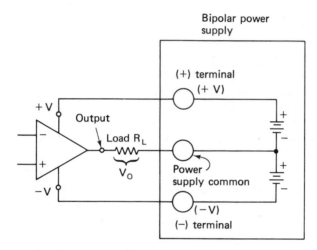

(a) Actual wiring from power supply to op amp

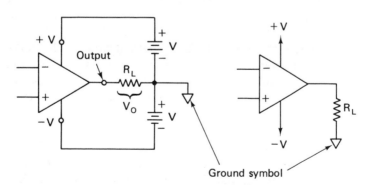

(b) Typical schematic representations of supplying power to an op amp

Figure 2-3 Wiring power and load to an op amp.

13

standard practice to show power common as a ground symbol on a schematic diagram. Use of the term "ground" or the ground symbol is a convention which indicates that all voltage measurements are made with respect to "ground"

The power supply in Fig. 2-3 is called a bipolar or split supply and has typical values of ± 15 V, ± 12 V, and ± 6 V. Special-purpose op amps may require nonsymmetrical supplies such as $+12$ V and -6 V, or even a single polarity supply such as $+30$ V and ground. Note that the ground is *not* wired to the op amp in Fig. 2-3. Currents returning to the supply from the op amp must return through external circuit elements such as the load resistor R_L. The maximum supply voltage that can be applied between $+V$ and $-V$ is typically 36 V or ± 18 V.

2-1.2 Output Terminal

In Fig. 2-3 the op amp's output terminal is connected to one side of the load resistor R_L. The other side of R_L is wired to ground. Output voltage V_o is measured with respect to ground. Since there is only one output terminal in an op amp, it is called a *single-ended output*. There is a limit to the current that can be drawn from the output terminal of an op amp, usually on the order of 5 to 10 mA. There are also limits on the output terminal's voltage levels; these limits are set by the supply voltages and by output transistors Q_{16} and Q_{11} in Fig. 2-1. (See also Appendix 1, "Output Voltage Swing as a Function of Supply Voltage.") These transistors need about 1 to 2 V from collector to emitter to ensure that they are acting as amplifiers and not as switches. Thus the output terminal can rise to within 2 V of $+V$ and drop to within 2 V of $-V$. The upper limit of V_o is called the *positive saturation voltage*, $+V_{sat}$, and the lower limit is called the *negative saturation voltage*, $-V_{sat}$. For example, with a supply voltage of ± 15 V, $+V_{sat} = +13$ V and $-V_{sat} = -13$ V. Therefore, V_o is restricted to a peak-to-peak swing of ± 13 V. Both current and voltage limits place a *minimum* value on the load resistance R_L of 2 kΩ. However, special-purpose op amps such as the CA3130 have MOS rather than bipolar output transistors. The output of a CA3130 can be brought to within millivolts of either $+V$ or $-V$.

Some op amps, such as the 741, have internal circuitry that automatically limits current drawn from the output terminal. Even with a short circuit for R_L, output current is limited to about 25 mA, as noted in Appendix 1. This feature prevents destruction of the op amp in the event of a short circuit.

2-1.3 Input Terminals

In Fig. 2-4 there are two input terminals, labeled $-$ and $+$. They are called *differential input terminals* because output voltage V_o depends on the *difference* in voltage between them, E_d, and the gain of the amplifier, A_{OL}. As shown in Fig. 2-4(a), the output terminal is positive with respect to ground when the $(+)$ input is positive with respect to the $(-)$ input. When E_d is reversed in Fig. 2-4(b) to make the $(+)$ input negative with respect to the $(-)$ input, V_o becomes negative with respect to ground.

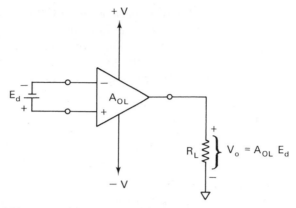

(a) V_o goes positive when the (+) input is positive with respect to the (−) input.

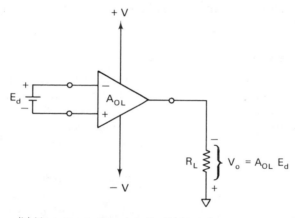

(b) V_o goes negative when the (+) input is negative with respect to the (−) input.

Figure 2-4 Polarity of V_o depends on the polarity of differential input voltage E_d.

We conclude from Fig. 2-4 that the polarity of the output terminal is the same as the polarity of (+) input terminal. Moreover, the polarity of the output terminal is opposite or inverted from the polarity of the (−) input terminal. For these reasons, the (−) input is designated the *inverting input* and the (+) input the *noninverting input* (see Appendix 1).

It is important to emphasize that the polarity of V_o depends only on the *difference* in voltage between inverting and noninverting inputs. This difference voltage can be found by

$$E_d = \text{voltage at the } (+) \text{ input} - \text{voltage at the } (-) \text{ input} \qquad (2\text{-}1)$$

Both input voltages are measured with respect to ground. The sign of E_d tells us (1) the polarity of the (+) input with respect to the (−) input and (2) the polarity of the output terminal with respect to ground. This equation holds if the inverting input is grounded, if the noninverting input is grounded, and even if both inputs are above or below ground potential. One other important characteristic of the input terminals is the high impedance between them and also between each input terminal and ground.

2-2   OPEN-LOOP VOLTAGE GAIN

2-2.1   Definition

Refer to Fig. 2-4. If differential input voltage E_d is small enough, output voltage V_o will be determined by both E_d and the *open-loop voltage gain*, A_{OL}. A_{OL} is called open-loop voltage gain because possible feedback connections from output terminal to input terminals are left open. Accordingly, V_o would be ideally expressed by the simple relationship

$$\text{output voltage} = \text{differential input voltage} \times \text{open loop gain}$$
$$V_o = E_d \times A_{OL} \tag{2-2}$$

2-2.2   Differential Input Voltage, E_d

The value of A_{OL} is extremely large, often 200,000 or more. Recall from Section 2-1.2 that V_o can never exceed positive or negative saturation voltages $+V_{sat}$ and $-V_{sat}$. For a ±15-V supply, saturation voltages would be about ±13 V. Thus, for the op amp to act as an amplifier, E_d must be limited to a maximum voltage of ±65 μV. This conclusion is reached from Eq. (2-2).

$$E_{d \text{ max}} = \frac{+V_{sat}}{A_{OL}} = \frac{13 \text{ V}}{200,000} = 65 \ \mu\text{V}$$

$$-E_{d \text{ max}} = \frac{-V_{sat}}{A_{OL}} = \frac{-13 \text{ V}}{200,000} = -65 \ \mu\text{V}$$

In the laboratory or shop it is difficult to measure 65 μV, because induced noise, 60-Hz hum, and leakage currents on the typical test setup can easily generate a millivolt (1000 μV). Furthermore, it is difficult and inconvenient to measure very high gains. The op amp also has tiny internal unbalances that *act* as a small voltage that offsets, and even may exceed, E_d. This *offset voltage* is discussed in Chapter 9.

2-2.3   Conclusions

There are three conclusions to be drawn from these brief comments. First, V_o in the circuit of Fig. 2-4 either will be at one of the limits $+V_{sat}$ or $-V_{sat}$ or will be oscillating between these limits. Don't be disturbed, because this behavior is what a high-gain amplifier usually does. Second, to maintain V_o between these limits we must go to a feedback type of circuit that forces V_o to depend on stable, precision elements such as resistors and signal generators rather than A_{OL} and E_d.

The last and most important conclusion is that *if E_d is so small that we cannot measure it easily, then for all practical purposes E_d equals* 0 V. This conclusion is by no means trivial. We will repeatedly use the fact that $E_d \approx 0$ *if V_o lies between the saturation voltages.* To state this in another way, the (−) input will always be at about the same potential as the (+) input with respect to ground if V_o lies between saturation voltages. Without learning any more about the op amp, it is possible to understand basic comparator applications. In a comparator application, the op amp performs not as an amplifier but as a device that tells when an unknown voltage is below, above, or just equal to a known reference voltage. Before introducing the comparator in the next section, an example is given to illustrate ideas presented thus far.

Example 2-1

In   Fig.   2-4, $+V = 15$ V, $-V = -15$ V, $+V_{sat} = +13$ V, $-V_{sat} = -13$ V, and gain $A_{OL} = 200,000$. Assuming ideal conditions, find the magnitude and polarity of V_o for each of the following input voltages, given with respect to ground.

	Voltage at (−) input	Voltage at (+) input
(a)	−10 μV	−15 μV
(b)	−10 μV	+15 μV
(c)	−10 μV	−5 μV
(d)	+1.000001 V	+1.000000 V
(e)	+5 mV	0 V
(f)	0 V	+5 mV

Solution. The polarity of V_o is the same as the polarity of the (+) input with respect to the (−) input. The (+) input is more negative than the (−) input in (a), (d), and (e). This is shown by Eq. (2-1), and therefore V_o will go negative. From Eq. (2-2), the magnitude of V_o is A_{OL} times the difference, E_d, between voltages at the (+) and (−) inputs. But if $A_{OL} \times E_d$ exceeds $+V$ or $-V$, then V_o must stop at $+V_{sat}$ or $-V_{sat}$ as in (e) and (f). Calculations are summarized as follows.

	E_d [using Eq. (2-1)]	Polarity of (+) input with respect to (−) input	Theoretical V_o [from Eq. (2-2)]	Polarity of output terminal with respect to ground
(a)	−5 μV	−	5 μV × 200,000 = −1.0 V	−
(b)	25 μV	+	25 μV × 200,000 = 5.0 V	+
(c)	5 μV	+	5 μV × 200,000 = 1.0 V	+
(d)	−1 μV	−	1 μV × 200,000 = −0.2 V	−
(e)	−5 mV	−	−13 V = −V_{sat}	−
(f)	5 mV	+	13 V = +V_{sat}	+

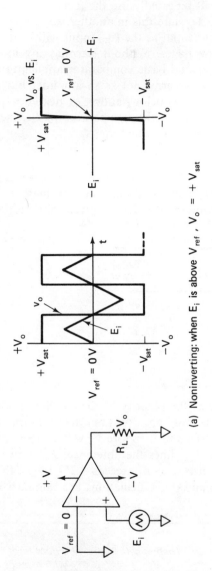

(a) Noninverting: when E_i is above V_{ref}, $V_o = +V_{sat}$

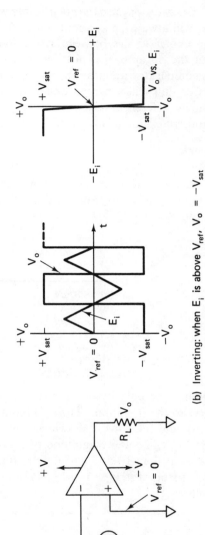

(b) Inverting: when E_i is above V_{ref}, $V_o = -V_{sat}$

Figure 2-5 Zero-crossing detectors, noninverting in (a) and inverting in (b).

2-3 ZERO-CROSSING DETECTORS

2-3.1 Noninverting Zero-Crossing Detector

The op amp in Fig. 2-5(a) operates as a comparator. Its $(+)$ input compares voltage E_i with a reference voltage of 0 V ($V_{ref} = 0$ V). When E_i is above V_{ref}, V_o equals $+V_{sat}$. This is because the voltage at the $(+)$ input is more positive than the voltage at the $(-)$ input. Therefore, the sign of E_d in Eq. (2-1) is positive. Consequently, V_o is positive, from Eq. (2-2).

The polarity of V_o tells *if* E_i is above or below V_{ref}. The *transition* of V_o tells *when* E_i crossed the reference and in what direction. For example, when V_o makes a positive-going transition from $-V_{sat}$ to $+V_{sat}$, it indicates that E_i just crossed 0 in the positive direction.

2-3.2 Inverting Zero-Crossing Detector

The op amp's $(-)$ input in Fig. 2-5(b) compares E_i with a reference voltage of 0 V ($V_{ref} = 0$ V). This circuit is an *inverting zero-crossing detector*. The waveshapes of V_o versus time and V_o versus E_i can be explained by the following summary:

1. If E_i is above V_{ref}, V_o equals $-V_{sat}$.
2. When E_i crosses the reference going positive, V_o makes a negative-going transition from $+V_{sat}$ to $-V_{sat}$.

2-4 POSITIVE- AND NEGATIVE-VOLTAGE-LEVEL DETECTORS

2-4.1 Positive-Level Detectors

In Fig. 2-6 a positive reference voltage V_{ref} is applied to one of the op amp's inputs. This means that the op amp is set up as a comparator to detect a positive voltage. If the voltage to be sensed, E_i, is applied to the op amp's $(+)$ input, the result is a *noninverting positive-level detector*. Its operation is shown by the waveshapes in Fig. 2-6(a). When E_i is above V_{ref}, V_o equals $+V_{sat}$. When E_i is below V_{ref}, V_o equals $-V_{sat}$.

If E_i is applied to the inverting input as in Fig. 2-6(b), the circuit is an inverting positive-level detector. Its operation can be summarized by the statement: When E_i is above V_{ref}, V_o equals $-V_{sat}$. This circuit action can be seen more clearly by observing the plot of E_i and V_{ref} versus time in Fig. 2-6(b).

2-4.2 Negative-Level Detectors

Figure 2-7(a) is a *noninverting negative-level detector*. This circuit detects when input signal E_i crosses the negative voltage $-V_{ref}$. When E_i is above $-V_{ref}$, V_o equals $+V_{sat}$. When E_i is below $-V_{ref}$, $V_o = -V_{sat}$.

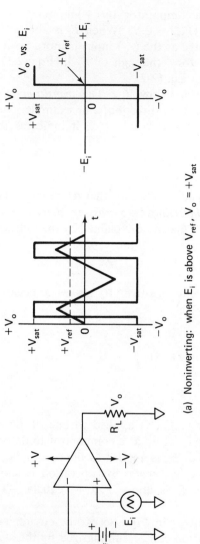

(a) Noninverting: when E_i is above V_{ref}, $V_o = +V_{sat}$

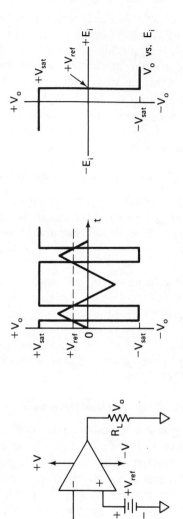

(b) Inverting: when E_i is above V_{ref}, $V_o = -V_{sat}$

Figure 2-6 Positive-voltage-level detector, noninverting in (a) and inverting in (b).

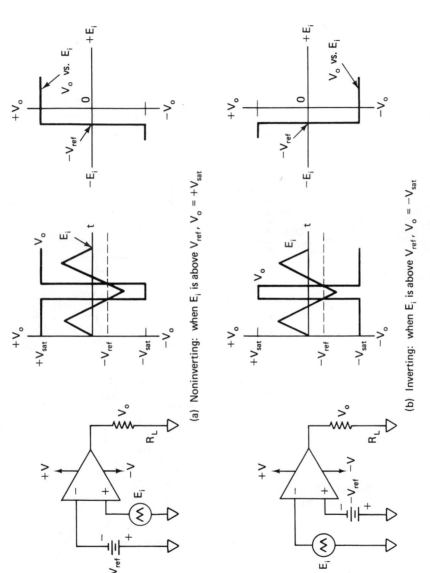

(a) Noninverting: when E_i is above V_{ref}, $V_o = +V_{sat}$

(b) Inverting: when E_i is above V_{ref}, $V_o = -V_{sat}$

Figure 2-7 Negative-voltage-level detector, noninverting in (a) and inverting in (b).

21

The circuit of Fig. 2-7(b) is an *inverting negative-level detector*. When E_i is above $-V_{ref}$, V_o equals $-V_{sat}$, and when E_i is below $-V_{ref}$, V_o equals $+V_{sat}$.

2-5 TYPICAL APPLICATIONS OF VOLTAGE-LEVEL DETECTORS

2-5.1 Sound-Activated Switch

Figure 2-8(a) shows how to make an adjustable reference voltage. Two 15-kΩ resistors and a 100-Ω potentiometer are connected in series as a voltage divider across the bipolar power supply. Normally, $+V$ and $-V$ are well regu-

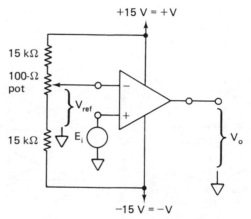

(a) A variable reference voltage can be obtained
     by using the op amp's bipolar supply along
     with a voltage-divider network

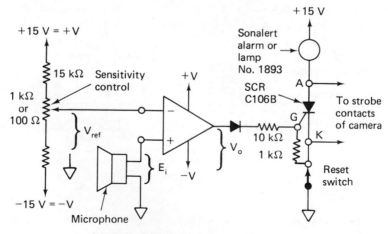

(b) Voltage-level detector as a sound switch

Figure 2-8 A sound-activated switch is made by connecting the output of a noninverting voltage-level detector to an alarm circuit.

lated and provide a stable reference voltage via the divider. V_{ref} can be varied from $+50$ mV to -50 mV by adjusting the 100-Ω pot.

A practical application that uses the level detector of Fig. 2-8(a) is the sound-activated switch shown in Fig. 2-8(b). Signal source E_i is a microphone and an alarm circuit is connected to the output. The procedure to arm the sound switch is as follows:

1. Open the reset switch to turn off both SCR and alarm.
2. In a quiet environment, adjust the sensitivity control until V_o just swings to $-V_{sat}$.
3. Close the reset switch. The alarm should remain off.

Any noise signal will now generate an ac voltage and be picked up by the microphone as an input. The first positive swing of E_i above V_{ref} drives V_o to $+V_{sat}$. The diode now conducts a current pulse of about 1 mA into the gate (G) of the silicon-controlled rectifier (SCR). Normally, the SCR's anode, A, and cathode, K, terminals act like an open switch. However, the gate current pulse makes the SCR turn on, and now the anode and cathode terminals act like a closed switch. The audible or visual alarm is now activated. Furthermore, the alarm stays on because once an SCR has been turned on, it stays on until its anode–cathode circuit is opened.

The circuit of Fig. 2-8(b) can be modified to photograph high-speed events such as a bullet penetrating a glass bulb. Some cameras have mechanical switch contacts that close to activate a stroboscopic flash. To build this sound-activated flash circuit, remove the alarm and connect anode and cathode terminals to the strobe input in place of the camera switch. Turn off the room lights. Open the camera shutter and fire the rifle at the glass bulb. The rifle's sound activates the switch. The strobe does the work of apparently stopping the bullet in midair. Close the shutter. The position of the bullet in relation to the bulb in the picture can be adjusted experimentally by moving the microphone closer or farther away from the rifle.

2-5.2 Light Column Voltmeter

A light column voltmeter displays a column of light whose height is proportional to voltage. Manufacturers of audio and medical equipment may replace analog meter panels with light column voltmeters because they are easier to read at a distance.

A light column voltmeter is constructed in the circuit of Fig. 2-9. R_{cal} is adjusted so that 1 mA flows through the equal resistor divider network R_1 to R_{10}. Ten separate reference voltages are established in 1-V steps from 1 V to 10 V.

When $E_i = 0$ V or is less than 1 V, the outputs of all op amps are at $-V_{sat}$. The silicon diodes protect the light-emitting diodes against excessive reverse bias voltage. When E_i is increased to a value between 1 V and 2 V only, the output of op amp No. 1 goes positive to light LED1. Note that the op amp's

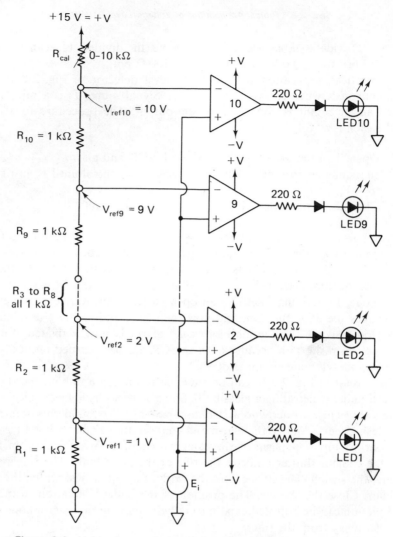

Figure 2-9 Light column voltmeter. Reference voltages to each op amp are set in steps of 1 V. As E_i is increased from 1 V to 10 V, LED1 through LED10 light in sequence.

output current is automatically limited by the op amp to its short-circuit value of about 20 mA. (The 220-Ω output resistors can be omitted if the op amps have short-circuit rather than foldback current characteristics.)

As E_i is increased, the LEDS light in numerical order. This circuit can also be built using five 747 dual op amps or two and one-half 4741 quad op amps.

2-5.3 Smoke Detector

Another practical application of a voltage-level detector is a smoke detector, as shown in Fig. 2-10. The lamp and photoconductive cell are mounted in an enclosed chamber that admits smoke but not external light. The photoconductor

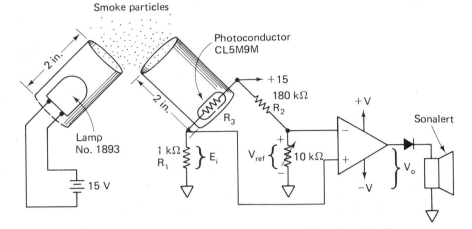

Figure 2-10 With no smoke present the 10-kΩ sensitivity control is adjusted until the alarm stops. Light reflected off any smoke particles causes the alarm to sound.

is a light-sensitive resistor. In the absence of smoke, very little light strikes the photoconductor and its resistance stays at some high value, typically several hundred kilohms. The 10-kΩ sensitivity control is adjusted until the alarm turns off.

Any smoke entering the chamber causes light to reflect off the smoke particles and strike the photo conductor. This, in turn, causes the photoconductor's resistance to decrease and the voltage across R_1 to increase. As E_i increases above V_{ref}, V_o switches from $-V_{sat}$ to $+V_{sat}$, causing the alarm to sound. The alarm circuit of Fig. 2-10 does not include an SCR. Therefore, when the smoke particles leave the chamber, the photoconductor's resistance increases and the alarm turns off. If you want the alarm to stay on, use the SCR alarm circuit shown in Fig. 2-8(b).

2-6 SIGNAL PROCESSING WITH VOLTAGE-LEVEL DETECTORS

2-6.1 Introduction

Armed with only the knowledge gained thus far, you can make a sine-to-square wave converter, duty-cycle controller, and pulse-width modulator out of the versatile op amp.

2-6.2 Sine-to-Square Wave Converter

The zero-crossing detector of Fig. 2-5 will convert the output of a sine-wave generator such as a variable-frequency audio oscillator into a variable-frequency square wave. If E_i is a sine, triangular, or wave of any other shape that is symmetrical around zero, the zero-crossing detector's output will be square. The frequency of E_i should be below 100 Hz, for reasons that are explained in Chapter 10.

2-6.3 Duty-Cycle Controller

The duty cycle, D, of a repetitive pulse signal is defined as the ratio of the *on* or *high* time (d) of the pulse to the period T of the signal. This ratio is usually multiplied by 100 to express duty cycle in percent. Mathematically, we have

$$\text{duty cycle } D = \frac{\text{high time } d}{\text{period } T} \times 100 \tag{2-3}$$

Period T is defined in Fig. 2-11 as the time interval between any two corresponding points on a wave. If E_i is a triangular wave, the duty cycle can be analyzed

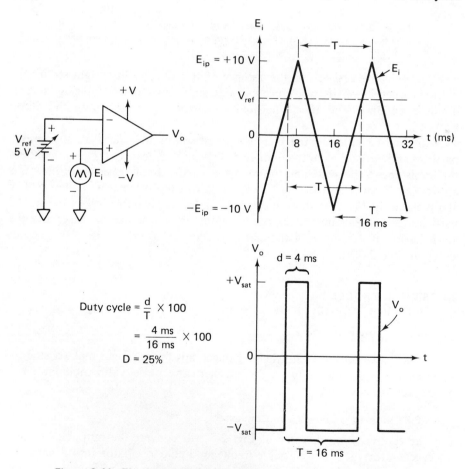

Figure 2-11   The duty cycle of a positive-level noninverting level detector can be controlled by varying V_{ref} and holding the amplitude of E_i constant.

or designed if you know the peak value of E_i, E_{ip}, and V_{ref} from

$$D = \frac{E_{ip} - V_{ref}}{2E_{ip}} \times 100 \qquad (2-4)$$

Example 2-2

Calculate the duty cycle in Fig. 2-11 if $E_{ip} = 10$ V and $V_{ref} = 5$ V.
Solution. From Eq. (2-4),

$$D = \frac{10 \text{ V} - 5 \text{ V}}{20 \text{ V}} \times 100 = 25\%$$

The ability to control duty cycle leads directly to the operating principles of a pulse-width modulator.

2-6.4 Pulse-Width Modulator

One way of transmitting information or data from one point to another is to send a *carrier wave*. The carrier is so named because it carries data. If the carrier wave is a pulse train of constant frequency and amplitude, there are three ways of loading data onto the carrier. The data can change or *modulate* either the carrier's amplitude, frequency, or pulse width. The resulting process is called *amplitude modulation, frequency modulation*, or *pulse-width modulation*, respectively, depending on which characteristic of the carrier is modulated.

A noninverting comparator is chosen to demonstrate the action of a pulse-width modulator [see Figs. 2-5(a) to 2-7(a)]. E_i is now a triangular carrier wave to be held constant in amplitude and frequency. Since frequency is constant, the period T of the wave is also constant ($T = 1/f$). E_i is applied to the $(+)$ input and has a peak value E_{ip}.

V_{ref} is applied to the $(-)$ input and is now considered to be the data signal. Let V_{ref} be positive and equal to $E_{ip}/2$, as shown in Fig. 2-12(a). V_o will exhibit a duty cycle of 25%, from Eq. (2-4). If data signal V_{ref} is reduced to 0 V, V_o will have a duty cycle of 50% in Fig 2-11(b). Finally, if V_{ref} is negative and equal to $E_{ip}/2$, the duty cycle increases to 75% in Fig. 2-12(c).

Figure 2-12 shows that the duty cycle D, and consequently the pulse width d, are proportional to data signal V_{ref}. Pulse width d can be found from

$$d = DT \qquad (2-5)$$

Example 2-3

In Fig. 2-12(c), $E_{ip} = 10$ V, $T = 10$ ms, and $V_{ref} = -5$ V. Find (a) the duty cycle; (b) the pulse width.
Solution. (a) From Eq. (2-4),

$$D = \frac{10 \text{ V} - (-5 \text{ V})}{2 \times 10 \text{ V}} \times 100 = \frac{15 \text{ V}}{20 \text{ V}} \times 100 = 75\%$$

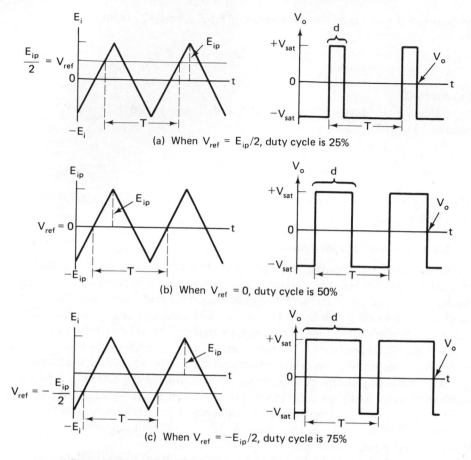

(a) When $V_{ref} = E_{ip}/2$, duty cycle is 25%

(b) When $V_{ref} = 0$, duty cycle is 50%

(c) When $V_{ref} = -E_{ip}/2$, duty cycle is 75%

Figure 2-12 A pulse-width modulator is made by applying a triangle wave (constant amplitude and frequency) to the (+) input of a comparator and varying a reference voltage at the (−) input.

(b) From Eq. (2-5),

$$d = \frac{75}{100}(10 \text{ ms}) = 7.5 \text{ ms}$$

Note the decimal form of the 75% duty cycle is used in the calculation.

2-7 PRACTICAL CONSIDERATIONS

When an op amp circuit is actually constructed, we learn that there are several differences between actual performance of a real op amp and the ideal situation presented thus far. First, the output voltage V_o does not change instantly from $-V_{sat}$ to $+V_{sat}$. It takes a measurable amount of time to make the change.

Second, when differential input voltage E_d is 0, V_o is not precisely 0. Finally, the op amp output voltage may even oscillate unpredictably, either just as E_d approaches 0 V or continuously. This points up the need to learn about a few more of the op amp's characteristics and how to minimize differences between ideal and actual performance. These topics are discussed in Chapters 9 and 10. More advanced applications will be studied in Chapter 4, but first we need to learn about the effect of feedback connections from the output to the ($-$) input; this is discussed in Chapter 3.

PROBLEMS

2-1. Name the five basic terminals of an op amp.

2-2. Name two types of op amp packages.

2-3. A 741 op amp is manufactured in a 14-pin dual-in-line package. What are the terminal numbers for the (a) inverting input; (b) noninverting input; (c) output?

2-4. A 741 op amp is connected to a ±15-V supply. What are the output terminal's operating limits under normal conditions with respect to (a) output voltage; (b) output current?

2-5. When the load resistor of an op amp is short-circuited, what is the op amp's (a) output voltage; (b) approximate output current?

2-6. Find the output voltage.

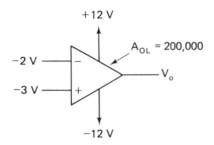

2-7. A triangular wave E_i is applied to the ($+$) input in Fig. 2-5(b). 0 V is applied to the ($-$) input. (a) Sketch V_o vs. time; (b) sketch V_o vs. E_i; (c) name the circuit.

2-8. If E_i were changed to a sine wave in Fig. 2-5(a) with the same frequency and peak amplitude, how would this affect the waveshapes in Fig. 2-5(a)?

2-9. Summarize the operation of a noninverting zero-crossing detector.

2-10. Summarize the operation of a noninverting voltage-level detector that has a reference voltage of $+3$ V.

2-11. To which input would you connect the reference voltage of an inverting voltage-level detector?

2-12. Sketch an adjustable (a) positive reference voltage, (b) negative reference voltage, and (c) bipolar reference voltage for a voltage-level detector. Use the supply voltage(s) and a single 100-kΩ potentiometer for each circuit.

2-13. Is the circuit of Fig. 2-8(b) an inverting or a noninverting voltage-level detector?

2-14. Classify as either an inverting or a noninverting voltage-level detector: (a) the circuit of Fig. 2-9; (b) the circuit of Fig. 2-10.

2-15. Sketch the circuit for a sine-to-square wave converter.

2-16. Calculate the duty cycle in Example 2-2 if V_{ref} is changed to (a) 0 V; (b) +5 V.

2-17. Plot pulse width in milliseconds versus V_{ref} as V_{ref} is varied from −10 V to +10 V in Example 2-3.

inverting and
noninverting amplifiers

3

3-0 INTRODUCTION

This chapter uses the op amp in one of its most important applications—making an amplifier. An *amplifier* is a circuit that receives a signal at its input and delivers an undistorted larger version of the signal at its output. All circuits in this chapter have one feature in common: An external feedback resistor is connected between the output terminal and ($-$) input terminal. This type of circuit is called a *negative feedback circuit.*

There are many advantages obtained with negative feedback, all based on the fact that circuit performance no longer depends on the open-loop gain of the op amp, A_{OL}. By adding the feedback resistor, we form a loop from output to ($-$) input. The resulting circuit now has a *closed-loop gain* or *amplifier gain*, A_{CL}, which is independent of A_{OL}.

As will be shown, closed-loop gain, A_{CL}, depends only on external resistors. For best results 1% resistors should be used, and A_{CL} will be known within about 1%. Note that adding external resistors does not change open-loop gain A_{OL}. A_{OL} still varies from op amp to op amp. So adding negative feedback will allow us to ignore changes in A_{OL} as long as A_{OL} is large. We begin with the inverting amplifier to show that A_{CL} depends simply on the ratio of two resistors.

3-1 THE INVERTING AMPLIFIER

3-1.1 Introduction

The circuit of Fig. 3-1 is one of the most widely used op amp circuits. It is an amplifier whose closed-loop gain from E_i to V_o is set by R_f and R_i. It can amplify ac or dc signals. To understand how this circuit operates, we make two realistic simplifying assumptions that were introduced in Chapter 2.

1. The voltage E_d between (+) and (−) inputs is essentially 0.
2. The current drawn by either the (+) or the (−) input terminal is negligible.

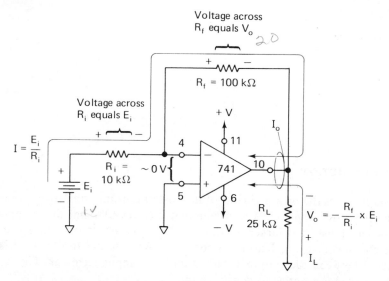

Figure 3-1 Positive voltage applied to the (−) input of an inverting amplifier. Numbers next to the small circles are pin connections for the 14-pin dual-in-line package.

3-1.2 Positive Voltage Applied to the Inverting Input

In Fig. 3-1, positive voltage E_i is applied through input resistor R_i to the op amp's (−) input. Negative feedback is provided by feedback resistor R_f. The voltage between (+) and (−) inputs is essentially equal to 0 V. Therefore, the (−) input terminal is also at 0 V, so ground potential is at the (−) input. For this reason, the (−) input is said to be at *virtual* ground.

Since one side of R_i is at E_i and the other is at 0 V, the voltage drop across R_i is E_i. The current I through R_i is found from Ohm's Law:

$$I = \frac{E_i}{R_i} \tag{3-1a}$$

R_i includes the resistance of the signal generator. This point is discussed further in Section 3-5.2.

All of the input current I flows through R_f, since a negligible amount is drawn by the $(-)$ input terminal. Note that the current through R_f is set by R_i and E_i; not by R_f, V_o, or the op amp.

The voltage drop across R_f is simply $I(R_f)$, or

$$V_{R_f} = I \times R_f = \frac{E_i}{R_i}R_f \qquad (3\text{-}1b)$$

But as shown in Fig. 3-1, one side of R_f and one side of load R_L are connected. The voltage from this connection to ground is V_o. The other sides of R_f and of R_L are at ground potential. Therefore, V_o equals V_{R_f} (the voltage across R_f). To obtain the polarity of V_o, note that the left side of R_f is at ground potential. The current direction established by E_i forces the right side of R_f to go negative. Therefore, V_o is negative when E_i is positive. Now, equating V_o with V_{R_f} and adding a minus sign to signify that V_o goes negative when E_i goes positive, we have

$$V_o = -E_i\frac{R_f}{R_i} \qquad (3\text{-}2a)$$

Now, introducing the definition that the closed-loop gain of the amplifier is A_{CL}, we rewrite Eq. (3-2a) as

$$A_{CL} = \frac{V_o}{E_i} = -\frac{R_f}{R_i} \qquad (3\text{-}2b)$$

The minus sign in Eq. (3-2b) shows that the polarity of the output V_o is inverted with respect to E_i. For this reason, the circuit of Fig. 3-1 is called an *inverting amplifier*.

3-1.3 Load and Output Currents

The load current I_L that flows through R_L is determined only by R_L and V_o and is furnished from the op amp's output terminal. Thus $I_L = V_o/R_L$. The current I through R_f must also be furnished by the output terminal. Therefore, the op amp output current I_o is

$$I_o = I + I_L \qquad (3\text{-}3)$$

The maximum value of I_o is set by op amp; it is usually between 5 and 10 mA.

Example 3-1

For Fig. 3-1, let $R_f = 100$ kΩ, $R_i = 10$ kΩ, and $E_i = 1$ V. Calculate (a) I; (b) V_o; (c) A_{CL}.

Solution. (a) From Eq. (3-1a),

$$I = \frac{E_i}{R_i} = \frac{1 \text{ V}}{10 \text{ k}\Omega} = 0.1 \text{ mA}$$

(b) From Eq. (3-2a),

$$V_o = -\frac{R_f}{R_i} \times E_i = -\frac{100 \text{ k}\Omega}{10 \text{ k}\Omega}(1 \text{ V}) = -10 \text{ V}$$

(c) Using Eq. (3-2b), we obtain

$$A_{CL} = -\frac{R_f}{R_i} = -\frac{100 \text{ k}\Omega}{10 \text{ k}\Omega} = -10$$

This answer may be checked by taking the ratio of V_o to E_i:

$$A_v = \frac{V_o}{E_i} = \frac{-10 \text{ V}}{1 \text{ V}} = -10$$

Example 3-2

Using the values given in Example 3-1 and $R_L = 25 \text{ k}\Omega$, determine (a) I_L and (b) the total current into the output pin of the op amp.

Solution. (a) Using the value of V_o calculated in Example 3-1, we obtain

$$I_L = \frac{V_o}{R_L} = \frac{10 \text{ V}}{25 \text{ k}\Omega} = 0.4 \text{ mA}$$

The direction of current is shown in Fig. 3-1.

(b) Using Eq. (3-3) and the value of I from Example 3-1, we obtain

$$I_o = I + I_L = 0.1 \text{ mA} + 0.4 \text{ mA} = 0.5 \text{ mA}$$

The input resistance seen by E_i is R_i. One of the reasons for using the op amp is its high input resistance. In order to keep input resistance of the *circuit* high, R_i should be equal to or greater than 10 kΩ.

3-1.4 Negative Voltage Applied to the Inverting Input

Figure 3-2 shows a negative voltage, E_i, applied via R_i to the inverting input. All the principles and equations of Sections 3-1.1 to 3-1.3 still apply. The only difference between Figs. 3-1 and 3-2 is the direction of the currents. Reversing the polarity of the input voltage, E_i reverses the direction of all currents and the voltage polarities. Now the output of the amplifier will go positive when E_i goes negative.

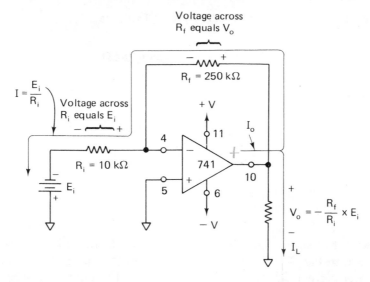

Figure 3-2 Negative voltage applied to the $(-)$ input of an inverting amplifier.

Example 3-3

For Fig. 3-2, let $R_f = 250\ \text{k}\Omega$, $R_i = 10\ \text{k}\Omega$, and $E_i = -0.5\ \text{V}$. Using Eq. (3-2a), calculate (a) I; (b) the voltage across R_f; (c) V_o.

Solution. (a) From Eq. (3-1a),

$$I = \frac{E_i}{R_i} = \frac{0.5\ \text{V}}{10\ \text{k}\Omega} = 50\ \mu\text{A} = 0.05\ \text{mA}$$

(b) From Eq. (3-1b),

$$V_{R_f} = I \times R_f$$
$$= (50\ \mu\text{A})(250\ \text{k}\Omega) = 12.5\ \text{V}$$

(c) From Eq. (3-2a),

$$V_o = -\frac{R_f}{R_i} \times E_i = -\frac{250\ \text{k}\Omega}{10\ \text{k}\Omega}(-0.5\ \text{V}) = +12.5\ \text{V}$$

Thus the magnitude of the output voltage does equal the voltage across R_f, and $A_{CL} = -25$.

Example 3-4

Using the values in Example 3-3, determine (a) R_L for a load current of 2 mA; (b) I_o; (c) the circuit's input resistance.

Solution. (a) Using Ohm's Law and V_o from Example 3-3,

$$R_L = \frac{V_o}{I_L} = \frac{12.5\text{ V}}{2\text{ mA}} = 6.25\text{ k}\Omega$$

(b) From Eq. (3-3) and Example 3-3,

$$I_o = I + I_L = 0.05\text{ mA} + 2\text{ mA} = 2.05\text{ mA}$$

(c) The circuit's input resistance, or the resistance seen by E_i is $R_i = 10\text{ k}\Omega$.

3-1.5 AC Voltage Applied to the Inverting Input

Figure 3-3 shows an ac signal voltage E_i applied to the inverting input. For the positive half-signal, the voltage polarities and the direction of currents are the same as in Fig. 3-1. For the negative half-signal voltage, the polarities and direction of currents are the same as in Fig. 3-2. The output waveform is the negative (or 180° out of phase) of the input wave as shown in Fig. 3-3. That is, when E_i is positive, V_o is negative; and vice versa. The equations developed in Section 3-1.2 are applicable to Fig. 3-3 for ac voltages.

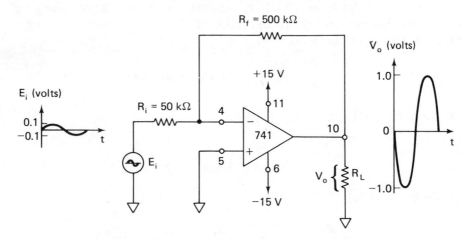

Figure 3-3 Inverting amplifier with ac input signal.

Example 3-5

For the circuit of Fig. 3-3, $R_f = 500\text{ k}\Omega$, and $R_i = 50\text{ k}\Omega$, calculate the voltage gain A_{CL}. See also Problem 3-19 for a plot of V_o vs. E_i.
Solution. From Eq. (3-2b),

$$A_{CL} = -\frac{R_f}{R_i} = \frac{-500\text{ k}\Omega}{50\text{ k}\Omega} = -10$$

Example 3-6

If the peak input voltage in Example 3-5 is 0.1 V, determine the peak output voltage.

Solution. Using Eq. (3-2b), we obtain

$$\text{peak } V_o = \frac{-R_f}{R_f} \times E_i = A_{CL}E_i = (-10)(0.1 \text{ V}) = -1.0 \text{ V}$$

The frequency of the output and input signals is the same.

3-2 INVERTING ADDER AND AUDIO MIXER

3-2.1 Inverting Adder

In the circuit of Fig. 3-4, V_o equals the sum of the input voltages with polarity reversed. Expressed mathematically,

$$V_0 = -(E_1 + E_2 + E_3) \tag{3-4}$$

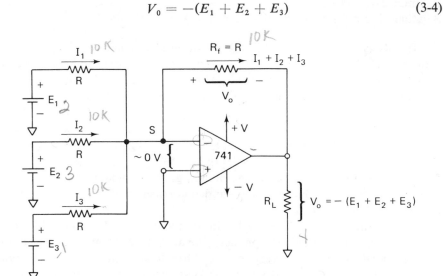

Figure 3-4 Inverting adder, $R = 10 \text{ k}\Omega$.

Circuit operation is explained by noting that the summing point S and the $(-)$ input are at ground potential. Current I_1 is set by E_1 and R, I_2 by E_2 and R, and I_3 by E_3 and R. Expressed mathematically,

$$I_1 = \frac{E_1}{R}, \qquad I_2 = \frac{E_2}{R}, \qquad I_3 = \frac{E_3}{R} \tag{3-5}$$

Since the $(-)$ input draws negligible current, I_1, I_2 and I_3 all flow through R_f. That is, the sum of the input currents flows through R_f and sets up a voltage drop across R_f equal to V_o, or

$$V_o = -(I_1 + I_2 + I_3)R_f$$

Substituting for the currents from Eq. (3-5) and substituting R for R_f, we obtain Eq. (3-4):

$$V_o = -\left(\frac{E_1}{R} + \frac{E_2}{R} + \frac{E_3}{R}\right)R = -(E_1 + E_2 + E_3)$$

Example 3-7

In Fig. 3-4, $E_1 = 2$ V, $E_2 = 3$ V, $E_3 = 1$ V, and all resistors are 10 kΩ. Evaluate V_o.
Solution. From Eq. (3-4), $V_o = -(2\text{ V} + 3\text{ V} + 1\text{ V}) = -6$ V.

Example 3-8

If the polarity of E_3 is reversed in Fig. 3-4 but the values are the same as in Example 3-7, find V_o.
Solution. From Eq. (3-4), $V_o = -(2\text{ V} + 3\text{ V} - 1\text{ V}) = -4$ V.

If only two input signals E_1 and E_2 are needed, simply replace E_3 with a short circuit to ground. If four signals must be added, simply add another equal resistor R between the fourth signal and summing point S. Equation (3-4) can be changed to include any number of input voltages.

3-2.2 Audio Mixer

In the adder of Fig. 3-4, all the input currents flow through feedback resistor R_f. This means that I_1 does not affect I_2 or I_3. More generally, the input currents do not affect one another because each sees ground potential at the summing node. Therefore, the input currents—and consequently the input voltages E_1, E_2, and E_3—do *not* interact.

This feature is especially desirable in an audio mixer. For example, let E_1, E_2, and E_3 be replaced by microphones. The ac voltages from each microphone will be added or mixed at every instant. Then if one microphone is carrying guitar music, it will not come out of a second microphone facing the singer. If a 100-kΩ volume control is installed between each microphone and associated input resistor, their relative volumes can be adjusted and added. A weak singer can then be heard above a very loud guitar.

3-3 INVERTING ADDER WITH GAIN

The three-input inverting adder in Fig. 3-4 is modified in Fig. 3-5 so that each input voltage can be multipled by a constant voltage gain and the results added. Just as in the adder, each input current is set by its input voltage and input resistance.

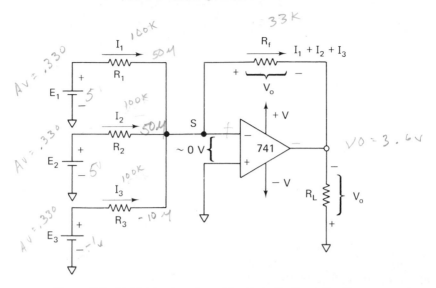

Figure 3-5 Multiplying inverting adder, $R_f > R_1, R_2$, and/or R_3.

$$I_1 = \frac{E_1}{R_1}, \qquad I_2 = \frac{E_2}{R_2}, \qquad I_3 = \frac{E_3}{R_3} \qquad (3\text{-}6)$$

Similarly, all input currents add together in R_f to generate an output voltage equal to R_f times the current sum, or

$$V_o = -(I_1 + I_2 + I_3)R_f = -\left(E_1\frac{R_f}{R_1} + E_2\frac{R_f}{R_2} + E_3\frac{R_f}{R_3}\right) \qquad (3\text{-}7)$$

Equation (3-7) shows that the gain for each input may be adjusted individually by choosing the desired ratio between R_f and each corresponding input resistor.

Example 3-9

In Fig. 3-5, $R_f = 100$ kΩ, $R_1 = 10$ kΩ, $R_2 = 20$ kΩ, and $R_3 = 50$ kΩ. Find (a) the *magnitude* of voltage gain applied to each input voltage and (b) the output voltage if $E_1 = E_2 = 0.1$ V and $E_3 = -0.1$ V.

Solution. (a) From Eq. (3-7), we can find the closed-loop gain A_{CL} for each input. For E_1,

$$|A_{CL_1}| = \frac{R_f}{R_1} = \frac{100 \text{ k}\Omega}{10 \text{ k}\Omega} = 10$$

For E_2,

$$|A_{CL_2}| = \frac{R_f}{R_2} = \frac{100 \text{ k}\Omega}{20 \text{ k}\Omega} = 5$$

For E_3,

$$|A_{CL_3}| = \frac{R_f}{R_3} = \frac{100 \text{ k}\Omega}{50 \text{ k}\Omega} = 2$$

(b) From Eq. (3-7),

$$V_o = -[0.1(10) + 0.1(5) + (-0.1)2]$$
$$= -(1.0 + 0.5 - 0.2) = -1.3 \text{ V}$$

3-4 INVERTING AVERAGING AMPLIFIER

An *averaging amplifier* gives an output voltage proportional to the average of all the input voltages. If there are three input voltages, the averager should add the input voltages and divide the sum by 3. The averager is the same circuit arrangement as the inverting adder in Fig. 3-4 or the inverting adder with gain in Fig. 3-5. The difference is that the input resistors are made equal to some convenient value R and the feedback resistor is made equal to R divided by the number of inputs. Let n equal the number of inputs. Then for a three-input averager, $n = 3$ and $R_f = R/3$. Proof is found by substituting into Eq. (3-7), for $R_f = R/3$ and $R_1 = R_2 = R_3 = R$ to show that

$$V_o = -\left(\frac{E_1 + E_2 + E_3}{n}\right) \tag{3-8}$$

Example 3-10

In Fig. 3-5, $R_1 = R_2 = R_3 = R = 100 \text{ k}\Omega$ and $R_f = 100 \text{ k}\Omega/3 = 33 \text{ k}\Omega$. If $E_1 = +5$ V, $E_2 = +5$ V, and $E_3 = -1$ V, find V_o.

Solution. Since $R_f = R/3$, the amplifier is an averager, and from Eq. (3-8) with $n = 3$, we have

$$V_o = -\left[\frac{5 \text{ V} + 5 \text{ V} + (-1 \text{ V})}{3}\right] = -\frac{9 \text{ V}}{3} = -3 \text{ V}$$

3-5 VOLTAGE FOLLOWER

3-5.1 Introduction

The circuit of Fig. 3-6 is called a *voltage follower*, but it is also referred to as a *source follower, unity-gain amplifier buffer amplifier*, or *isolation amplifier*. The input voltage, E_i, is applied directly to the (+) input. Since the voltage between (+) and (−) pins of the op amp can be considered 0,

$$V_o = E_i \tag{3-9a}$$

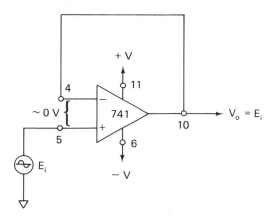

Figure 3-6 Voltage follower.

Note that the output voltage equals the input voltage in both magnitude and sign. Therefore, as the name of the circuit implies, the output voltage *follows* the input or source voltage. The voltage gain is 1 (or unity), as shown by

$$A_{CL} = \frac{V_o}{E_i} = 1 \tag{3-9b}$$

Example 3-11

For Fig. 3-7(a), determine (a) V_o; (b) I_L; (c) I_o.

Solution. (a) From Eq. (3-9a),

$$V_o = E_i = 4 \text{ V}$$

(b) From Ohm's Law,

$$I_L = \frac{V_o}{R_L} = \frac{4 \text{ V}}{10 \text{ k}\Omega} = 0.4 \text{ mA}$$

(c) From Eq. (3-3),

$$I_o = I + I_L$$

This circuit is still a negative-feedback amplifier because there is a connection between output and (−) input. Remember that it is negative feedback that forces E_d to be 0 V. But $I \approx 0$, since input terminals of op amps draw negligible current; therefore,

$$I_o = 0 + 0.4 \text{ mA} = 0.4 \text{ mA}$$

If E_i were reversed, the polarity of V_o and the direction of currents would be reversed, as shown in Fig. 3-7(b).

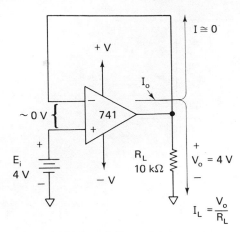

(a) Voltage follower for a
    positive input voltage

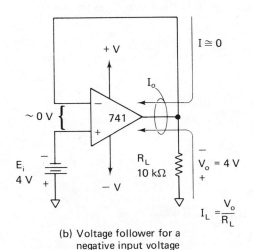

(b) Voltage follower for a
    negative input voltage

Figure 3-7 Circuits for Example 3-11.

3-5.2 Using the Voltage Follower

A question that arises quite often is: Why bother to use an amplifier with a gain of 1? The answer is best seen if we compare a voltage follower with an inverting amplifier. In this example, we are not primarily concerned with the polarity of voltage gain but rather with the input loading effect.

The voltage follower is used because its input resistance is high (many megohms). Therefore, it draws negligible current from a signal source. For example, in Fig. 3-8(a) the signal source has an open circuit or generator voltage, E_{gen}, of 1.0 V. The generator's internal resistance is 90 kΩ. Since the input terminal of the op amp draws negligible current, the voltage drop across R_{int}

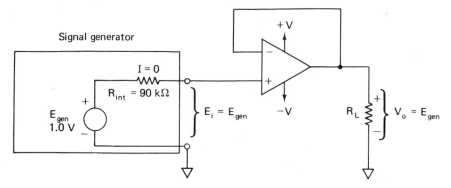

(a) Essentially no current is drawn from E_{gen}. The
output terminal of the op amp can supply up
to 5 mA with a voltage held constant at E_{gen}

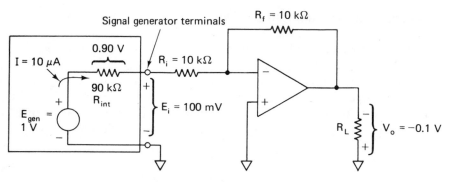

(b) E_{gen} divides between its own internal
resistance and amplifier input resistance

Figure 3-8   Comparison of loading effect between inverting and noninverting
amplifiers on a high-resistance source.

is 0 V. The terminal voltage E_i of the signal source becomes the input voltage
to the amplifier and equals E_{gen}. Thus

$$V_o = E_i = E_{gen}$$

Now let us consider the same signal source connected to an inverting
amplifier whose gain is -1 [see Fig. 3-8(b)]. As stated in Section 3-1.3, the input
resistance to an inverting amplifier is R_i. This causes the generator voltage E_{gen}
to divide between R_{int} and R_i. Using the voltage division law to find the generator terminal voltage E_i yields

$$E_i = \frac{R_i}{R_{int} + R_i} \times E_{gen} = \frac{10 \text{ k}\Omega}{10 \text{ k}\Omega + 90 \text{ k}\Omega} \times (1.0 \text{ V}) = 0.1 \text{ V}$$

Thus it is this 0.1 V that becomes the input voltage to the inverting amplifier. If the inverting amplifier has a gain of only -1, the output voltage V_o is -0.1 V.

In conclusion, if a high-impedance sourse is connected to an inverting amplifier, the voltage gain from V_o to E_{gen} is not set by R_f and R_i as given in Eq. (3-2b). The actual gain must include R_{int}, as

$$\frac{V_o}{E_{gen}} = -\frac{R_f}{R_i + R_{int}}$$

If you must amplify and invert a signal source from a high-impedance source and wish to draw no signal current, first *buffer* the source with a voltage follower. Then feed the follower's output into an inverter. We now turn to a circuit that will amplify and buffer but *not* invert a signal source, the noninverting amplifier.

3-6 NONINVERTING AMPLIFIER

Figure 3-9 is a noninverting amplifier; that is, the output voltage, V_o, is the same polarity as the input voltage, E_i. The input resistance of the inverting amplifier (Section 3-1) is R_i, but the input resistance of the noninverting amplifier is extremely large, typically exceeding 100 MΩ. Since there is practically 0 voltage between the $(+)$ and $(-)$ pins of the op amp, both pins are at the same potential E_i. Therefore, E_i appears across R_1. E_i causes current I to flow as given by

$$I = \frac{E_i}{R_1} \tag{3-10a}$$

The direction of I depends on the polarity of E_i. Compare Fig. 3-9(a) and (b). The input current to the op amp's $(-)$ terminal is negligible. Therefore, I flows through R_f and the voltage drop across R_f is represented by V_{R_f} and expressed as

$$V_{R_f} = I(R_f) = \frac{R_f}{R_1} \times E_i \tag{3-10b}$$

Equations (3-10a) and (3-10b) are similar to Eqs. (3-1a) and (3-1b).

The output voltage V_0 is found by adding the voltage drop across R_1, which is E_i, to the voltage across R_f, which is V_{R_f}:

$$V_o = E_i + \frac{R_f}{R_1} E_i$$

or

$$V_o = \left(1 + \frac{R_f}{R_1}\right) E_i \tag{3-11a}$$

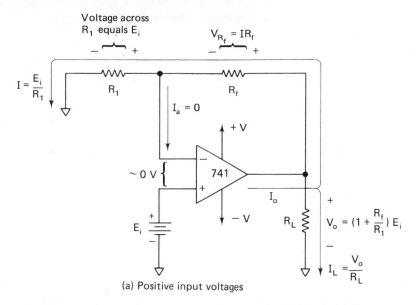

(a) Positive input voltages

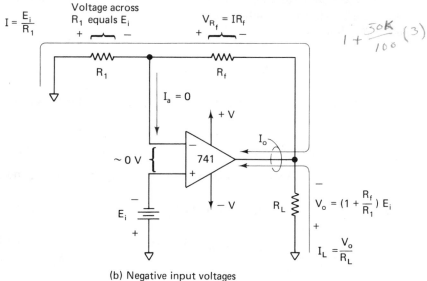

$$1 + \frac{50\text{K}}{100} \quad (3)$$

(b) Negative input voltages

Figure 3-9 Voltage polarities and direction of currents for noninverting amplifiers.

Rearranging Eq. (3-11a) to express voltage gain, we get

$$A_{CL} = \frac{V_o}{E_i} = 1 + \frac{R_f}{R_1} \tag{3-11b}$$

Equation (3-11b) shows that the voltage gain of a noninverting amplifier equals the *magnitude* of the gain of an inverting amplifier (R_f/R_1) plus 1.

The load current I_L is given by V_o/R_L and therefore depends only on V_o and R_L. I_o, the current drawn from the output pin of the op amp, is given by Eq. (3-3).

Example 3-12

For the circuit of Fig. 3-9(a), let $R_1 = 5$ kΩ, $R_f = 20$ kΩ, and $E_i = 2$ V. Calculate (a) V_o and (b) A_{CL}. See also Problem 3-19 for a plot of V_o vs. E_i.
Solution. (a) From Eq. (3-11a),

$$V_o = \left(1 + \frac{20 \text{ k}\Omega}{5 \text{ k}\Omega}\right)(2 \text{ V}) = 10 \text{ V}$$

(b) From Eq. (3-11b),

$$A_{CL} = \frac{V_o}{E_i} = \frac{10 \text{ V}}{2 \text{ V}} = 5$$

or

$$A_{CL} = 1 + \frac{R_F}{R_1} = 1 + \frac{20 \text{ k}\Omega}{5 \text{ k}\Omega} = 1 + 4 = 5$$

Example 3-13

Using the circuit values of Example 3-12 and $R_L = 5$ kΩ, calculate (a) the load current I_L; (b) the output op amp current I_o.
Solution. (a) Since $V_o = 10$ V in Example 3-12,

$$I_L = \frac{V_o}{R_L} = \frac{10 \text{ V}}{5 \text{ k}\Omega} = 2 \text{ mA}$$

(b) Applying Eq. (3-3) and the value of $I = 2$ V/5 k$\Omega = 0.4$ mA, we obtain

$$I_o = I + I_L = 0.4 \text{ mA} + 2 \text{ mA} = 2.4 \text{ mA}$$

Example 3-14

The circuit of Fig. 3-9 is to be designed for $A_{CL} = 16$, $E_i = 0.2$ V, and $R_1 = 2$ kΩ; calculate R_f.
Solution. (a) Rearranging Eq. (3-11b), we have

$$\frac{R_f}{R_1} = A_{CL} - 1 = 16 - 1 = 15$$

Then

$$R_f = 15R_1 = 15(2 \text{ k}\Omega) = 30 \text{ k}\Omega$$

3-7 NONINVERTING ADDER

3-7.1 Two-Input Noninverting Adder

A two-input noninverting adder is shown in Fig. 3-10. The voltage at the (+) input E_i is found from the nodal equation:

$$\frac{E_i - E_1}{R} + \frac{E_i - E_2}{R} = 0 \qquad \text{so that} \qquad E_i = \frac{E_1 + E_2}{2}$$

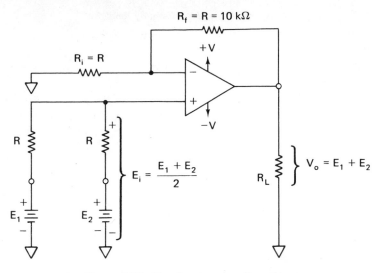

Figure 3-10 Two-input noninverting adder.

E_i is then multiplied by $+2$ to give V_o:

$$V_o = E_1 + E_2 \qquad (3\text{-}12)$$

If the signals sources E_1 and E_2 do not have negligible resistance with respect to resistor R, each should be buffered with a voltage follower.

3-7.2 N-input Noninverting Adder

If more than two input signals are to be added, we make all resistors equal except the feedback resistor R_f. For the three-input noninverting adder in Fig. 3-11,

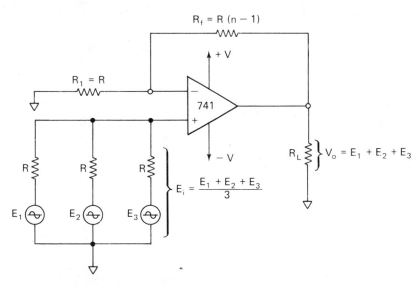

Figure 3-11 Three-input ($n = 3$) noninverting adder; $R = 10$ kΩ.

47

R_f is made equal to

$$R_f = (n - 1)R \qquad (3\text{-}13)$$

where n is the number of inputs. Now E_i is the sum of the input voltages divided by the number of inputs (the average input voltage). The gain of the amplifier is then set to equal the number of inputs. Therefore, V_o simply adds the input voltages.

Example 3-15

In Fig. 3-11, $E_1 = E_2 = 2$ V and $E_3 = -1$ V. If $R_1 = R = 10$ kΩ, find (a) n; (b) R_f; (c) V_o.

Solution. (a) $n = 3$; (b) by Eq. (3-13), $R_f = (3-1)10$ kΩ = 20 kΩ; (c) $V_o = E_1 + E_2 + E_3 = 2$ V + 2 V − 1 V = 3 V.

3-8 SINGLE-SUPPLY OPERATION

With the exception of a few special-purpose op amps (such as the CA3140 BiMOS op amp) it is impossible to make a full-range dc amplifier with a single-polarity power supply and an op amp. This is because the output terminal and both input terminals of an op amp *cannot* be brought closer than about 2 V to either supply voltage. So if the $-V$ supply terminal is at ground potential, neither output nor inputs can be brought lower than $+2$ V.

However, an ac amplifier can be constructed by holding the op amp's input and output terminals at some convenient dc voltage that is usually half of the single supply voltage. For example, in Fig. 3-12 the equal 220-kΩ resistors R_B

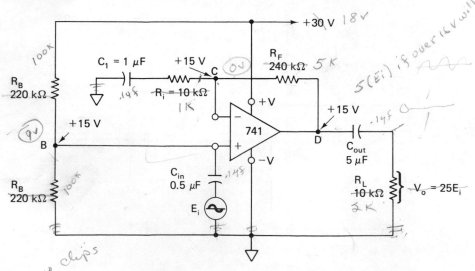

Figure 3-12 Construction of an ac amplifier with an op amp and a single-polarity power supply.

divide the 30-V supply voltage in half to set point B at $+15$ V with respect to ground. Point C must go to $+15$ V because the op amp's differential input voltage E_d equals 0 V. No dc current flows through R_i, and consequently R_f, because of capacitor C_1. Therefore, point D is at $+15$ V.

Only the ac component of signal source E_i is coupled through C_{in} to the op amp's $(+)$ input. E_i sees an input resistance equal to the parallel combination of the 220-kΩ resistors, or 110 kΩ. R_i and R_f form a noninverting amplifier for ac signals with a gain of $(R_F + R_i)/R_i = 25$. C_o blocks the 15-V bias voltage at point D and transmits only the amplified ac signals to load R_L.

3-9 DIFFERENCE BETWEEN MEASURED AND CALCULATED VALUES

If the circuits of this chapter are tested, there may be differences between measured and calculated values of output voltage. These differences will be due to the fact that unavoidable op amp limitations have not been accounted for. Mention of such limitations in this chapter would divert attention from and needlessly complicate understanding of circuit operation. These limitations are, however, covered thoroughly in Chapters 9 and 10.

PROBLEMS

3-1. What type of feedback is applied to an op amp when an external component is connected between the output terminal and the inverting input?

3-2. If the open-loop gain is very large, does the closed-loop gain depend on the external components or the op amp?

3-3. What two assumptions have been used to analyze the circuits of this chapter?

3-4. Repeat Example 3-1 with $R_i = 50$ kΩ.

3-5. In Fig. 3-1, if $R_L = 10$ kΩ, determine (a) I_L; (b) I_o. $E_i = 1$ V.

3-6. If $R_i = 50$ kΩ in Fig. 3-1, what is the input resistance as seen by E_i?

3-7. In Example 3-3, if $E_i = 0.05$ V, calculate (a) I; (b) V_{R_i}; (c) V_o.

3-8. In Fig. 3-2, if $E_i = 0.4$ V and $R_L = 5$ kΩ, determine (a) I_L; (b) I_o.

3-9. If $R_f = 100$ kΩ and $R_i = 20$ kΩ, will the closed-loop gain be the same for both dc and ac input signals in a noninverting amplifier?

3-10. For Fig. 3-3, if $\pm V_{sat} = \pm 15$ V, what peak-to-peak value of E_i begins to cause the output to saturate?

3-11. Calculate V_o for Fig. 3-4 if (a) $E_1 = -2$ V, $E_2 = -1$ V, and $E_3 = +0.5$ V; (b) $E_1 = 2$ V, $E_2 = -3$ V, and $E_3 = 1$ V. All resistors equal 20 kΩ.

3-12. Using the values in Example 3-7, determine (a) the load current and (b) the current into the output terminal of the op amp. $R_L = 10$ kΩ.

3-13. Repeat Example 3-9 for $R_f = 50$ kΩ.

3-14. If $E_1 = -8$ V, $E_2 = +2$ V, and $E_3 = 0$ in Example 3-10, find V_o.

3-15. In Fig. 3-7(a), if $E_i = +10$ V, calculate (a) V_o; (b) I_L; (c) I_o.

3-16. Give two other names for the voltage-follower circuits of Fig. 3-6.

3-17. Calculate V_o in Fig. 3-8(b) if R_{int} is changed to 40 kΩ.

3-18. Let R_f be changed to 10 kΩ in Examples 3-12 and 3-13. Find the new values for (a) A_{CL}; (b) I_L; (c) I_o.

3-19. An input signal E_i is varied between -10 V and $+10$ V as it is applied to both an inverting amplifier and a noninverting amplifier. If the gain of the inverting amplifier is -5 and the gain of the noninverting amplifier is $+5$, plot V_o vs. E_i for each amplifier on the same graph. Assume that $\pm V_{sat} = \pm 15$ V.

3-20. Design an inverting amplifier with a gain of -10 and an input resistance of 10 kΩ.

3-21. Design a noninverting amplifier with a gain of $+10$. Let R_i equal 10 kΩ.

3-22. Design a four-input (a) inverting adder; (b) noninverting adder.

3-23. Design a noninverting ac amplifier with a gain of 10 that will work with a single supply voltage of 15 V.

comparators

4

4-0 INTRODUCTION

A comparator compares a signal voltage on one input with a reference voltage on the other input. Voltage-level-detector circuits were introduced in Chapter 2 to show how easy it is to use op amps to solve some types of signal comparison applications without the need to know much about the op amp itself. The general-purpose op amp was used as a substitute for ICs designed only for comparator applications.

Unfortunately, the general-purpose op amp's output voltage does not change very rapidly. Also, its output changes between limits fixed by the saturation voltages, $+V_{sat}$ and $-V_{sat}$ that are typically about ± 13 V. Therefore, the output cannot drive devices, such as TTL digital logic ICs, that require voltage levels between 0 and $+5$ V. These disadvantages are eliminated by an IC that has been specifically designed to act as a comparator. One such device is the 311 comparator.

Neither the general-purpose op amp nor the comparator can operate properly if noise is present at either input. To solve this problem, we will learn how the addition of *positive feedback* overcomes the noise problem. Note that positive feedback does not eliminate the noise but makes the op amp less responsive to it. These circuits will show how to make better voltage-level detectors and also build a foundation to understand square-wave generators (multivibrators) and single-pulse generators (one-shots) that are covered in Chapter 6.

4-1 EFFECT OF NOISE ON COMPARATOR CIRCUITS

Input signal E_i is applied to the $(-)$ input of a 301 op amp in Fig. 4-1 (the 301 is a general-purpose op amp). If no noise is present, the circuit operates as an inverting zero-crossing detector because $V_{ref} = 0$.

Noise voltage E_n is shown, for simplicity, as a square wave in series with E_i. To show the effect of noise voltage, the op amp's input signal voltage is drawn both with and without noise in Fig. 4-2. The waveshape of V_o vs. time shows clearly how the addition of noise causes false output signals. V_o should indicate only the crossings of E_i, *not* the crossings of E_i plus noise voltage.

If E_i approaches V_{ref} very slowly or actually hovers close to V_{ref}, V_o can either follow all the noise voltage oscillations or burst into high-frequency oscillation. These false crossings can be eliminated by *positive feedback*.

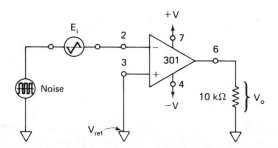

Figure 4-1   Inverting zero-crossing detector.

4-2 POSITIVE FEEDBACK

4-2.1 Introduction

Positive feedback is accomplished by taking a fraction of the output voltage V_o and applying it to the $(+)$ input. In Fig. 4-3(a), output voltage V_o divides between R_1 and R_2. A fraction of V_o is fed back to the $(+)$ input and creates a reference voltage that depends on V_o. The idea of a reference voltage was introduced in Chapter 2. We will now study positive feedback and how it can be used to eliminate false output changes due to noise.

4-2.2 Upper-Threshold Voltage

In Fig. 4-3(a), output voltage V_o divides between R_1 and R_2. A fraction of V_o is fed back to the $(+)$ input. When $V_o = +V_{sat}$, the fed-back voltage is called the *upper-threshold voltage* V_{UT}. V_{UT} is expressed from the voltage divider as

$$V_{UT} = \frac{R_2}{R_1 + R_2}(+V_{sat}) \tag{4-1}$$

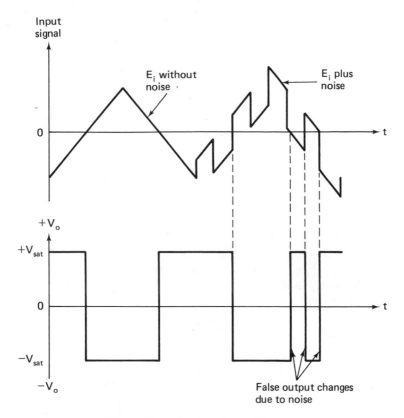

The addition of noise voltage at the
input causes false zero crossings.

Figure 4-2   Effect of noise on a zero-crossing detector.

For E_i values below V_{UT}, the voltage at the $(+)$ input is greater than the voltage
at the $(-)$ input. Therefore, V_o is locked at $+V_{sat}$.

If E_i is made slightly more positive than V_{UT}, the polarity of E_d, as shown,
reverses and V_o begins to drop in value. Now the fraction of V_o fed back to the
positive input is smaller, so E_d becomes larger. V_o then drops even faster and
is driven quickly to $-V_{sat}$. The circuit is then stable at the condition shown in
Fig. 4-3(b).

4-2.3   Lower-Threshold Voltage

When V_o is at $-V_{sat}$, the voltage fed back to the $(+)$ input is called *lower-threshold voltage* V_{LT} and is given by

$$V_{LT} = \frac{R_2}{R_1 + R_2}(-V_{sat}) \tag{4-2}$$

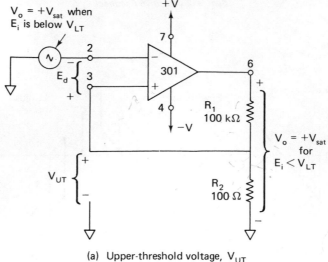

(a) Upper-threshold voltage, V_{UT}

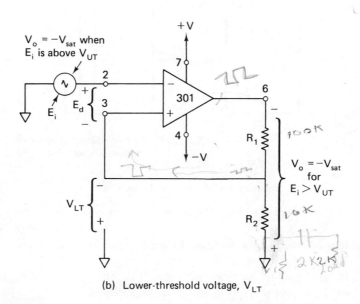

(b) Lower-threshold voltage, V_{LT}

Figure 4-3 R_1 and R_2 feed back a reference voltage from the output to the (+) input terminal.

Note that V_{LT} is negative with respect to ground. Therefore, V_o will stay at $-V_{sat}$ as long as E_i is above, or positive with respect to, V_{LT}. V_o will switch back to $+V_{sat}$ if E_i goes more negative than, or below, V_{LT}.

We conclude that positive feedback induces a snap action to switch V_o faster from one limit to the other. Once V_o begins to change, it causes a regenerative action that makes V_o change even faster. If the threshold voltages are larger than the peak noise voltages, positive feedback will eliminate false output transitions. This principle is investigated in the following examples.

Example 4-1

If $+V_{sat} = 14$ V in Fig. 4-3(a), find V_{UT}. ✓
Solution. By Eq. (4-1),

$$V_{UT} = \frac{100\,\Omega}{100,100\,\Omega}(14\text{ V}) \approx 14\text{ mV}$$

Example 4-2

If $-V_{sat} = -13$ V in Fig. 4-3(b), find V_{LT}. ✓
Solution. By Eq. (4-2),

$$V_{LT} = \frac{100\,\Omega}{100,100\,\Omega}(-13\text{ V}) \approx -13\text{ mV}$$

Example 4-3

In Fig. 4-4, E_i is a triangular wave applied to the $(-)$ input in Fig. 4-3(a).
Find the resultant output voltage.

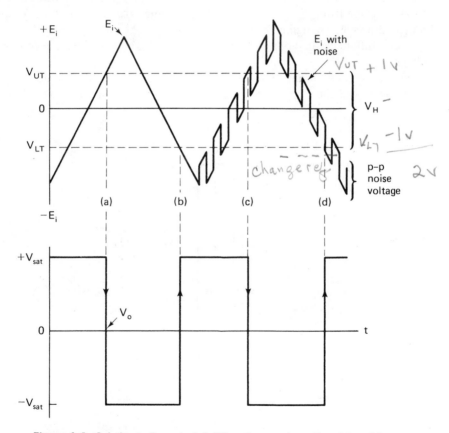

Figure 4-4   Solution to Example 4-3. When E_i goes above V_{UT} at time (c),
V_o goes to $-V_{sat}$. The peak-to-peak noise voltage would have to equal or
exceed V_H to pull E_i below V_{LT} and generate a false crossing. Thus V_H tells
us the margin against peak-to-peak noise voltage.

Solution. The dashed lines drawn on E_i in Fig. 4-4 locate V_{UT} and V_{LT}. At time $t = 0$, E_i is below V_{LT}, so V_o is at $+V_{sat}$ (as in Fig. 4-4). When E_i goes above V_{UT}, at times (a) and (c), V_o switches quickly to $-V_{sat}$. When E_i again goes below V_{LT}, at times (b) and (d), V_o switches quickly to $+V_{sat}$. Observe how positive feedback has eliminated the false crossings.

4-3 ZERO-CROSSING DETECTOR WITH HYSTERESIS

4-3.1 Defining Hysteresis

There is a standard technique of showing comparator performance on one graph instead of two graphs, as in Fig. 4-4. By plotting E_i on the horizontal axis and V_o on the vertical axis, we obtain the output–input voltage characteristic, as in Fig. 4-5. For E_i less than V_{LT}, $V_o = +V_{sat}$. The vertical line (a) shows V_o going from $+V_{sat}$ to $-V_{sat}$ as E_i becomes greater than V_{UT}. Vertical line (b) shows V_o changing from $-V_{sat}$ to $+V_{sat}$ when E_i becomes less than V_{LT}. The difference in voltage between V_{UT} and V_{LT} is called the *hysteresis voltage*, V_H.

Whenever any circuit changes from one state to a second state at some input signal and then reverts from the second to the first state at a *different*

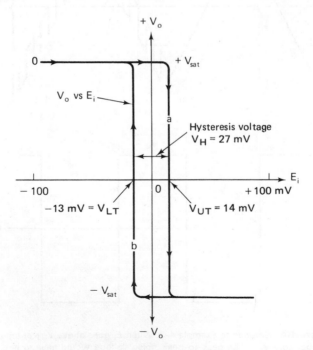

Figure 4-5 Plot of V_o vs. E_i illustrates the amount of hysteresis voltage in a comparator circuit.

input signal, the circuit is said to exhibit *hysteresis*. For the positive-feedback comparator, the difference in input signals is

$$V_H = V_{UT} - V_{LT} \tag{4-3}$$

For Examples 4-1 and 4-2, the hysteresis voltage is $14\,\text{mV} - (-13\,\text{mV}) = 27\,\text{mV}$.

If the hysteresis voltage is designed to be greater than the *peak-to-peak* noise voltage, there will be no false output crossings. Thus V_H tells us how much peak-to-peak noise the circuit can withstand.

4-3.2 Zero-Crossing Detector with Hysteresis as a Memory Element

If E_i has a value that lies between V_{LT} and V_{UT}, it is impossible to predict the value of V_o unless you already know the value of V_o. For example, suppose that you substitute ground for E_i ($E_i = 0$ V) in Fig. 4-3 and turn on the power. The op amp will go to *either* $+V_{sat}$ *or* $-V_{sat}$, depending on the inevitable presence of noise. If the op amp goes to $+V_{sat}$, E_i must then go above V_{UT} in order to change the output. If V_o had gone to $-V_{sat}$, then E_i would have to go below V_{LT} to change V_o.

Thus the comparator with hysteresis exhibits the property of *memory*. That is, if E_i lies between V_{UT} and V_{LT} (within the hysteresis voltage), the op amp remembers whether the last switching value of E_i was above V_{UT} or below V_{LT}.

4-4 VOLTAGE-LEVEL DETECTORS WITH HYSTERESIS

4-4.1 Introduction

In the zero-crossing detectors of Sections 4-2 and 4-3, the hysteresis voltage V_H is centered on the zero reference voltage V_{ref}. It is also desirable to have a collection of circuits that exhibit hysteresis about a center voltage that is either positive or negative. For example, an application may require a positive output, V_o, when an input E_i goes above an upper threshold voltage of $V_{UT} = 12$ V. Also, we may wish V_o to go negative when E_i goes below a lower threshold voltage of, for example, $V_{LT} = 8$ V. These requirements are summarized on the plot of V_o vs. E_i in Fig. 4-6. V_H is evaluated from Eq. (4-3) as

$$V_H = V_{UT} - V_{LT} = 12\,\text{V} - 8\,\text{V} = 4\,\text{V}$$

The hysteresis voltage V_H should be centered on the average of V_{UT} and V_{LT}. This average is called center voltage V_{ctr}, where

$$V_{ctr} = \frac{V_{UT} + V_{LT}}{2} = \frac{12\,\text{V} + 8\,\text{V}}{2} = 10\,\text{V}$$

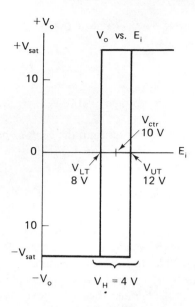

Figure 4-6 Positive-voltage-level detector. Hysteresis voltage V_H is symmetrical about the center voltage V_{ctr}. This voltage level detector is a noninverting type because V_o goes positive when E_i goes above V_{UT}.

When we try to build this type of voltage-level detector, it is desirable to have four features: (1) an adjustable resistor to set and refine the value of V_H; (2) a separate adjustable resistor to set the value of V_{ctr}; (3) the setting of V_H and V_{ctr} should *not* interact; and (4) the center voltage V_{ctr} should equal or be simply related to an external reference voltage V_{ref}. For the lowest possible parts count, the op amp's regulated supply voltage and a resistor network can be used for selecting V_{ref}.

Sections 4-4.2 and 4-4.3 deal with circuits that do not have all these features but are low in parts count and consequently cost. Section 4-5 presents a circuit that has all four features but at the cost of a higher parts count.

4-4.2 Noninverting Voltage-Level Detector with Hysteresis

The positive feedback resistor from output to (+) input indicates the presence of hysteresis in the circuit of Fig. 4-7. E_i is applied via R to the (+) input, so the circuit is noninverting. (Note that E_i must be a low-impedance source or the output of either a voltage follower or op amp amplifier.) The reference voltage V_{ref} is applied to the op amp's (−) input.

The upper- and lower-threshold voltages can be found from the following equations:

$$V_{UT} = V_{ref}\left(1 + \frac{1}{n}\right) - \frac{-V_{sat}}{n} \qquad (4\text{-}4a)$$

$$V_{LT} = V_{ref}\left(1 + \frac{1}{n}\right) - \frac{+V_{sat}}{n} \qquad (4\text{-}4b)$$

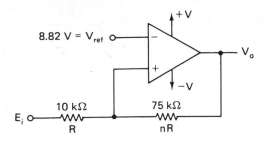

(a) The ratio of nR to R or n and V_{ref} determine V_{UT}, V_{LT}, V_H, and V_{ctr}

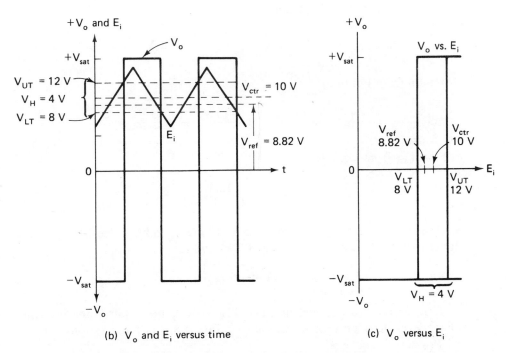

(b) V_o and E_i versus time

(c) V_o versus E_i

Figure 4-7 Noninverting voltage-level detector with hysteresis. Center voltage V_{ctr} and hysteresis voltage V_H cannot be adjusted independently since both depend on the ratio n.

Hysteresis voltage V_H is expressed by

$$V_H = V_{UT} - V_{LT} = \frac{(+V_{sat}) - (-V_{sat})}{n} \qquad (4\text{-}5)$$

In zero-crossing detectors, V_H is centered on the zero-volts reference. For the circuit of Fig. 4-7, V_H is *not* centered on V_{ref} but is symmetrical about the *average* value of V_{UT} and V_{LT}. This average value is called *center voltage* V_{ctr}

and is found from

$$V_{ctr} = \frac{V_{UT} + V_{LT}}{2} = V_{ref}\left(1 + \frac{1}{n}\right) \tag{4-6}$$

Compare the locations of V_{ctr} and V_{ref} in Figs. 4-6 and 4-7(c). Also compare Eqs. (4-5) and (4-6) to see that n appears in *both* equations. This means that any adjustment in resistor nR affects *both* V_{ctr} and V_H.

Example 4-4

Design the circuit of Fig. 4-7 to have $V_{UT} = 12$ V and $V_{LT} = 8$ V. Assume that $\pm V_{sat} = \pm 15$ V.
Solution. (a) From Eqs. (4-5) and (4-6),

$$V_H = 12\text{ V} - 8\text{ V} = 4\text{ V}, \qquad V_{ctr} = \frac{12\text{ V} + 8\text{ V}}{2} = 10\text{ V}$$

(b) Find n from Eq. (4-5):

$$n = \frac{+V_{sat} - (-V_{sat})}{V_H} = \frac{+15\text{ V} - (-15\text{ V})}{4} = 7.5$$

(c) Find V_{ref} from Eq. (4-6):

$$V_{ref} = \frac{V_{ctr}}{1 + 1/n} = \frac{10\text{ V}}{1 + 1/7.5} = 8.82\text{ V}$$

(d) Select $R = 10$ kΩ and $nR = 7.5 \times 10$ kΩ $= 75$ kΩ. The relationships between E_i and V_o are shown in Fig. 4-7(b) and (c).

4-4.3 Inverting Voltage-Level Detector with Hysteresis

If E_i and V_{ref} are interchanged in Fig. 4-7(a), the result is the inverting voltage-level detector with hysteresis (see Fig. 4-8). The expressions for V_{UT} and V_{LT} are

$$V_{UT} = \frac{n}{n+1}(V_{ref}) + \frac{+V_{sat}}{n+1} \tag{4-7a}$$

$$V_{LT} = \frac{n}{n+1}(V_{ref}) + \frac{-V_{sat}}{n+1} \tag{4-7b}$$

V_{ctr} and V_H are then found to be

$$V_{ctr} = \frac{V_{UT} + V_{LT}}{2} = \left(\frac{n}{n+1}\right)V_{ref} \tag{4-8}$$

$$V_H = V_{UT} - V_{LT} = \frac{+V_{sat} - (-V_{sat})}{n+1} \tag{4-9}$$

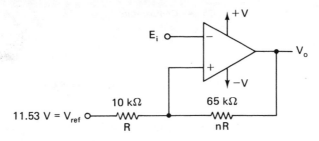

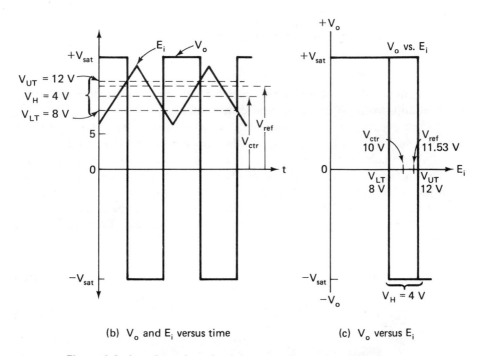

(a) The ratio of nR to R or n and V_{ref}
determine V_{UT}, V_{LT}, V_H, and V_{ctr}

(b) V_o and E_i versus time

(c) V_o versus E_i

Figure 4-8 Inverting voltage-level detector with hysteresis. Center voltage V_{ctr} and V_H cannot be adjusted independently since both depend on n.

Note that V_{ctr} and V_H both depend on n and therefore are *not* independently adjustable.

Example 4-5

Complete a design for Fig. 4-8 that has $V_{UT} = 12$ V and $V_{LT} = 8$ V. To make this example comparable with Example 4-5, assume that $\pm V_{sat} = \pm 15$ V. Therefore, $V_{ctr} = 10$ V and $V_H = 4$ V.

Solution. (a) Find n from Eq. (4-9):

$$n = \frac{+V_{sat} - (-V_{sat})}{V_H} - 1 = \frac{15 \text{ V} - (-15 \text{ V})}{4 \text{ V}} - 1 = 6.5$$

(b) Find V_{ref} from Eq. (4-8):

$$V_{ref} = \frac{n+1}{n}(V_{ctr}) = \frac{6.5+1}{6.5}(10) = 11.53 \text{ V}$$

(c) Choose $R = 10 \text{ k}\Omega$; therefore, resistor nR will be $6.5 \times 10 \text{ k}\Omega = 65 \text{ k}\Omega$. These circuit values and wave shapes are shown in Fig. 4-8.

4-5 VOLTAGE-LEVEL DETECTOR WITH INDEPENDENT ADJUSTMENT OF HYSTERESIS AND CENTER VOLTAGE

4-5.1 Introduction

The circuit of Fig. 4-9 is a noninverting voltage-level detector with independent adjustment of hysteresis and center voltage.

In this circuit, the center voltage V_{ctr} is determined by both resistor mR and the reference voltage V_{ref}. V_{ref} can be either supply voltage $+V$ or $-V$. Remember that the op amp's supply voltage is being used for a lower parts count. Hysteresis voltage V_H is determined by resistor nR. If resistor nR is adjustable, then V_H can be adjusted independently of V_{ctr}. Adjusting resistor mR adjusts V_{ctr} without affecting V_H. Note that the signal source, E_i, must be a low-impedance source. The key voltages are shown in Fig. 4-9 and are designed or evaluated from the following equations:

$$V_{UT} = -\frac{-V_{sat}}{n} - \frac{V_{ref}}{m} \tag{4-10a}$$

$$V_{LT} = \frac{-V_{ref}}{m} - \frac{+V_{sat}}{n} \tag{4-10b}$$

$$V_H = V_{UT} - V_{LT} = \frac{+V_{sat} - (-V_{sat})}{n} \tag{4-11}$$

$$V_{ctr} = \frac{V_{UT} + V_{LT}}{2} = -\frac{V_{ref}}{m} - \frac{+V_{sat} + (-V_{sat})}{2n} \tag{4-12a}$$

The general equation for V_{ctr} seems complex. However, if the magnitudes of $+V_{sat}$ and $-V_{sat}$ are nearly equal, then V_{ctr} is expressed simply by

$$V_{ctr} = -\frac{V_{ref}}{m} \tag{4-12b}$$

So V_{ctr} depends only on m, and V_H depends only on n.

The following example shows how easy it is to design a battery-charger control circuit.

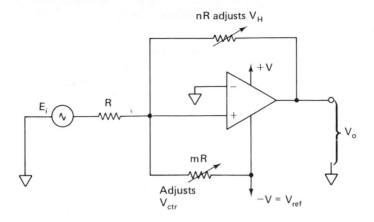

(a) Comparator with independent adjustments for hysteresis and reference voltage

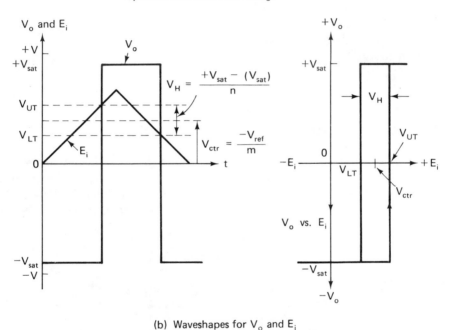

(b) Waveshapes for V_o and E_i

Figure 4-9 Resistor mR and supply voltage $-V$ establish the center voltage V_{ctr}. Resistor nR allows independent adjustment of the hysteresis voltage V_H, symmetrically around V_{ctr}.

4-5.2 Design Example:
Battery-Charger Control Circuit

Example 4-6

Assume that you want to monitor a 12–V battery. When the battery's voltage drops below 10.5 V, you want to connect it to a charger. When the battery voltage reaches 13.5 V, you want the charger to be disconnected. Therefore,

$V_{LT} = 10.5$ V and $V_{UT} = 13.5$ V. Let us use the $-V$ supply voltage for V_{ref} and assume that it equals -15.0 V. Further, let us assume that $\pm V_{sat} = \pm 13.0$ V. Find (a) V_H and V_{ctr}; (b) resistor mR; (c) resistor nR.

Solution. (a) From Eqs. (4-11) and (4-12),

$$V_H = V_{UT} - V_{LT} = 13.5 \text{ V} - 10.5 \text{ V} = 3.0 \text{ V}$$

$$V_{ctr} = \frac{V_{UT} + V_{LT}}{2} = \frac{13.5 \text{ V} + 10.5 \text{ V}}{2} = 12.0 \text{ V}$$

Note that the center voltage is the battery's nominal voltage.

(b) Arbitrarily choose resistor R to be a readily available value of $100 \text{ k}\Omega$. From Eq. (4-12b),

$$m = -\left(\frac{V_{ref}}{V_{ctr}}\right) = -\left(\frac{-15 \text{ V}}{12 \text{ V}}\right) = 1.25$$

Therefore, $mR = 1.25 \times 100 \text{ k}\Omega = 125 \text{ k}\Omega$

(c) From eq. (4-11),

$$n = \frac{+V_{sat} - (-V_{sat})}{V_H} = \frac{13 \text{ V} - (-13 \text{ V})}{3} = 8.66$$

Therefore, $nR = 866 \text{ k}\Omega$.

The final circuit is shown in Fig. 4-10. When E_i drops below 10.5 V, V_o goes negative, releasing the relay to its normally closed position. The relay's normally closed contacts (NC) connect the charger to battery E_i. Diode D_1 protects the transistor against excessive reverse bias when $V_o = -V_{sat}$. When

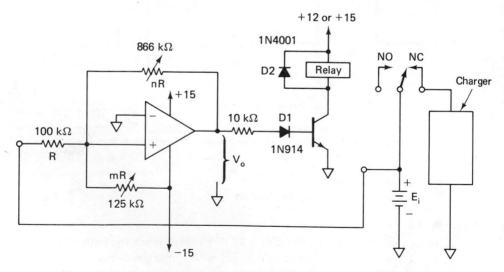

Figure 4-10 Battery-charger control for solution to Example 4-6. Adjust *mR* for $V_{ctr} = 12$ V in the test circuit of Fig. 4-9 and adjust *nR* for $V_H = 3$ V centered on V_{ctr}.

the battery charges to 13.5 V, V_o switches to $+V_{sat}$, which turns on the transistor and operates the relay. Its NC contacts open to disconnect the charger. Diode D_2 protects both op amp and transistor against transients developed by the relay's collapsing magnetic field.

One final note. Suppose that the application requires an inverting voltage-level detector with hysteresis. That is, V_o must go low when E_i goes above V_{UT} and V_o must go high when E_i drops below V_{LT}. For this application, do not change the circuit or design procedure for the noninverting voltage-level detectors, simply add an inverting amplifier to the output V_o.

4-6 CONTROLLING OUTPUT VOLTAGES OF VOLTAGE-LEVEL DETECTORS

One deficiency in the previous voltage-level detectors is that the output voltage switches between $\pm V_{sat}$. A 10-kΩ potentiometer can be connected between the op amp's output terminal and ground as a variable voltage divider. But any current drawn from its wiper changes its terminal voltage.

Three possible output-voltage waveforms are shown in Fig. 4-11. Assume that the switch is on position "(+) out." When the op amp's output voltage

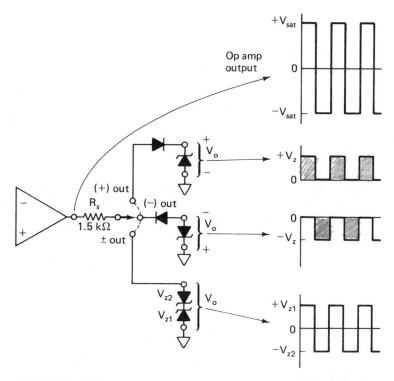

Figure 4-11   The output from a voltage-level detector can be changed from $\pm V_{sat}$ to a value determined by the zener diode.

is at $+V_{\text{sat}}$, the zener conducts producing the positive output voltage equal to the zener voltage V_z. When the op amp output voltage is at $-V_{\text{sat}}$, the blocking diode assures that the new output voltage will be zero. If the blocking diode is removed, the new output voltage equals the forward voltage of a zener, about -0.6 V.

When the switch is in position "($-$) out," the zener delivers a negative output of $-V_z$ when the op amp output is at $-V_{\text{sat}}$, and 0 V when the op amp's output voltage is $+V_{\text{sat}}$. For a symmetrical output, throw the switch to "$\pm$ out." When the op amp output voltage is at $+V_{\text{sat}}$, output V_o is positive at a value equal to ($V_{z1} + 0.6$ V from the forward-biased V_{z2}). When the op amp's output voltage is at $-V_{\text{sat}}$, V_o will equal $-(V_{z2} + 0.6$ V). The 0.6 V is from the forward-biased V_{z1}.

Load currents of up to 5 mA can be drawn from the zener outputs if R_s is calculated from

$$R_s \cong \frac{|V_{\text{sat}}| - V_z}{5\text{ mA}} \qquad (4\text{-}13)$$

For example, if $|V_{\text{sat}}| = 13$ V and $V_z = 5$ V, then $R_s = (13\text{ V} - 5\text{ V})/5\text{ mA} \cong 1.5$ kΩ. Feedback connections for hysteresis can now be taken from the zener output. The appropriate zener voltage now replaces $|+V_{\text{sat}}|$ or $|-V_{\text{sat}}|$ in the equations for V_{UT}, V_{LT}, V_{H}, and V_{ctr}.

4-7 IC PRECISION COMPARATOR, 111/311

4-7.1 Introduction

The 111 (military) or 311 (commercial) comparator is an IC that has been designed and optimized for superior performance in voltage-level-detector applications. A comparator should be fast. That is, its output should respond quickly to changes at its inputs. The 311 is much faster than the 741 or 301 but not as fast as the 710 and NE522 high-speed comparators. The subject of speed is discussed in Section 4-9, "Propagation Delay."

The 311 is an excellent choice for a comparator because of its versatility. Its output is designed *not* to bounce between $\pm V_{\text{sat}}$ but can be changed quite easily. As a matter of fact, if you are interfacing to a system with a different supply voltage, you simply connect the output to the new supply voltage via an appropriate resistor. We begin by examining operation of the output terminal.

4-7.2 Output Terminal Operation

A simplified model of the 311 in Fig. 4-12(a) shows that its output behaves like a switch Sw connected between output pin 7 and pin 1. Pin 7 can be wired to any voltage V^{++} with magnitudes up to 40 V more positive than the $-V$ supply terminal (pin 4). When ($+$) input pin 2 is more positive than ($-$) input pin 3, the 311's equivalent output switch is open. V_o is then determined by V^{++} and is $+5$ V.

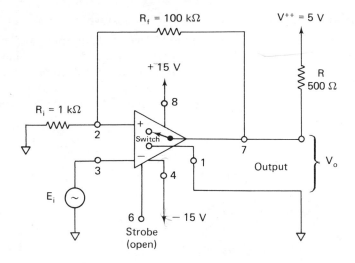

(a) 311 0-crossing detector with hysteresis

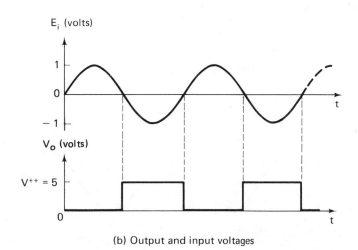

(b) Output and input voltages

Figure 4-12   Simplified model of the 311 comparator with input and output voltage waveforms.

When the $(+)$ input is less positive than (below) the $(-)$ input, the 311's equivalent output switch closes and extends the ground on pin 1 to output pin 7. R_f and R_i add about 50 mV of hysteresis to minimize noise effects so that pin 2 is essentially at 0 V. Waveshapes for V_o and E_i are shown in Fig. 4-12(b). V_o is 0 V (switch closed) for positive half-cycles of E_i. V_o is $+5$ V (switch open) for negative half-cycles of E_i. This is a typical interface circuit; that is, voltages may vary between levels of $+15$ V and -15 V, but V_o is restrained between $+5$ V and 0 V, which are typical digital signal levels. So the 311 can be used for converting analog voltage levels to digital voltage levels (interfacing).

4-7.3 Strobe Terminal Operation

The strobe terminal of the 311 is pin 6 (see also Appendix 3). This strobe feature allows the comparator output either to respond to input signals or to be independent of input signals. Fig. 4-13 uses the 311 comparator as a 0-crossing detector. A 10-kΩ resistor is connected to the strobe terminal. The other side of the resistor is connected to a switch. With the strobe switch open,

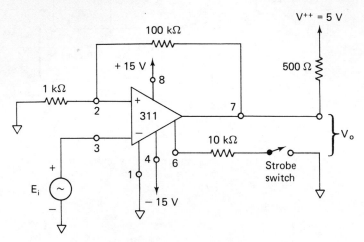

(a) 311 with strobe control

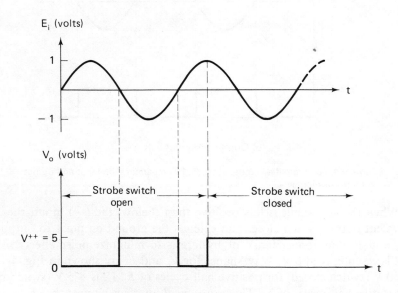

(b) When the strobe switch is closed, $V_o = V^{++}$

Figure 4-13 Operation of the strobe terminal.

the 311 operates normally. That is, the output voltage is at V^{++} for negative values of E_i and 0 for positive values of E_i. When the strobe switch is closed (connecting the 10 kΩ to ground), the output voltage goes to V^{++} regardless of the input signal. V_o will stay at V^{++} as long as the strobe switch is closed [see Fig. 4-13(b)]. The output is then independent of the inputs until the strobe switch is again opened.

The strobe feature is useful when a comparator is used to determine what type of signal is to be read out of a computer memory. The strobe switch is closed to ignore extraneous input signals that may occur up until the readout is due. Then during the readout time the switch is opened,˙and the 311 performs as a regular comparator. Current from the strobe terminal should be limited to about 3 mA. If the strobe feature is not to be used, the strobe terminal is left open or wired to $+V$ (see Appendix 3).

4-8 WINDOW DETECTOR

4-8.1 Introduction

The circuit of Fig. 4-14 is designed to monitor an input voltage and indicate when this voltage goes either above or below prescribed limits. For example, IC logic power supplies for TTL must be regulated to 5.0 V. If the supply voltage should exceed 5.5 V, the logic may be damaged, and if the supply voltage should drop below 4.5 V, the logic may exhibit marginal operation. Therefore, the limits for TTL power supplies are 4.5 V and 5.5 V. The power supply should be looking through a window whose limits are 4.5 V and 5.5 V, hence the name *window detector*. This circuit is sometimes called a *double-ended limit detector*.

In Fig. 4-14 input voltage E_i is connected to the $(-)$ input of comparator A and the $(+)$ input of comparator B. Upper limit V_{UT} is applied to the $(+)$ input of A, while lower limit V_{LT} is applied to the $(-)$ input of B. When E_i lies between V_{LT} and V_{UT}, the light/alarm is off. But when E_i drops below V_{LT} or goes above V_{UT}, the light/alarm goes on to signify that E_i is not between the prescribed limits.

4-8.2 Circuit Operation

Circuit operation is as follows. Assume that $E_i = 5$ V. Since E_i is greater than V_{LT} and less than V_{UT}, the output voltage of both comparators is at V^{++} because both output switches are open. The lamp/alarm is off. Next, assume that $E_i = 6.0$ V or $E_i > V_{UT}$. The input at pin 3 of A is more positive than at pin 2, so the A output is at the potential of pin 1 or ground. This ground lights the lamp, and $V_o = 0$ V. Now assume that E_i drops to 4.0 V or $E_i < V_{LT}$. The $(+)$ input of B is less than its $(-)$ input, so the B output goes to 0 V (the voltage at its pin 1). Once again this ground causes the lamp/alarm to light. Note that this application shows that output pins of the 311 can be connected together and the output is at V^{++} only when the output of each comparator is at V^{++}.

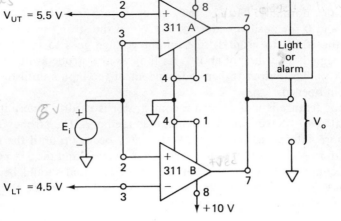

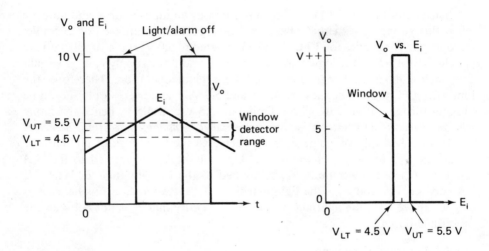

(a) Window detector circuit

(b) Waveshapes for window detector

Figure 4-14 Upper- and lower-threshold voltages are independently adjustable in the window detector circuit.

4-9 PROPAGATION DELAY

4-9.1 Definition

Suppose that a signal E_i is applied to the input of a comparator as in Fig. 4-15. There will be a measurable time interval for the signal to propagate through all the transistors within the comparator. After this time interval the output

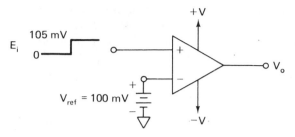

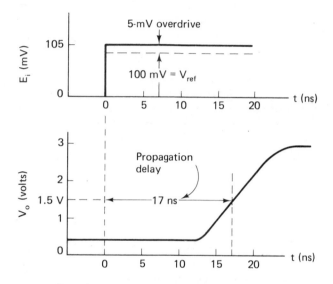

(a) Test circuit for propagation delay

(b) Propagation delay is the time interval between start of an input step voltage and the output rise to 1.5 V (NE522 comparator)

Figure 4-15 Propagation delay is measured by the test circuit in (a) and defined by the waveshapes in (b).

begins to change. This time interval is called *response time, transit time,* or *propagation delay.*

Before the signal is applied, the comparator is in saturation. This means that some of its internal transistors contain an excess amount of charge. It is the time required to clean out these charges that is primarily responsible for propagation delay.

4-9.2 Measurement of Propagation Delay

The comparator's depth of saturation depends directly on the amount of differential input voltage. The conventional method of comparing performance of one comparator with another is to first connect a $+100$-mV reference voltage to one input. In Fig. 4-15(a) the reference voltage is connected to the $(-)$ input. The other $(+)$ input is connected to 0 V. This forces all comparators, under test for propagation delay, into the same initial state of saturation. In Fig. 4-15(b) the outputs are shown at about 0.4 V before time 0.

A fast-rising signal voltage E_i is then applied to the $(+)$ input at time $t = 0$ in Fig. 4-15. If E_i if brought up to 100 mV, the comparator will be on the verge of switching but will *not* switch. However, if E_i is brought up quickly to 100 mV plus a small amount of *overdrive*, the overdrive signal will propagate through the comparator. After a propagation delay the output comes out of saturation and rises to a specified voltage. This voltage is typically 1.5 V.

As shown in Fig. 4-15(b), a 5-mV overdrive results in a propagation delay of 17 ns for a NE522 comparator. Increasing the overdrive to 100 mV will reduce propagation delay to 10 ns. Typical response times for the 311, 522, and 710 comparators, and the 301 op amp are:

Comparator	Response time for 5-mV overdrive (ns)	Response time for 20-mV overdrive (ns)
*311	170	100
522	17	15
710	40	20
301	> 10,000	> 10,000

*V^{++} = 5 V with a 500-Ω pull-up resistor.

PROBLEMS

4-1. Give a definition for a comparator circuit.

4-2. Suppose that noise is present in the input signal of a zero-crossing detector. What effect will the noise voltage have on the output voltage?

4-3. How do you recognize when positive feedback is present in the schematic of an op amp circuit?

4-4. If the value of R_2 is changed to 200 Ω in Fig. 4-3, what are the new values of V_{UT} and V_{LT} in Examples 4-1 and 4-2?

4-5. Find the hysteresis voltage in Problem 4-4.

4-6. (a) Plot V_o vs. E_i for Problem 4-4. (b) How much peak-to-peak noise voltage can this circuit withstand without generating false crossings in the output?

4-7. Redesign the circuit of Fig. 4-7 to have $V_{UT} = 10$ V and $V_{LT} = 8$ V.

4-8. Redesign the circuit of Fig. 4-8 to have $V_{UT} = 10$ V and $V_{LT} = 8$ V.

4-9. Redesign the battery charger of Example 4-6 to monitor a 9-V battery. Make $V_{UT} = 9.5$ V and $V_{LT} = 8.5$ V.

4-10. Assume that noise voltage is negligible in the zero-crossing detector of Fig. 4-1 and the saturation voltages are ± 15 V. Use Fig. 4-11 as guidance to provide output voltage levels of approximately ± 10 V.

4-11. Refer to the circuit of Fig. 4-12(a). What is the value of V_o if the strobe terminal is wired to $+V$ and (a) $E_i = 0.1$ V; (b) $E_i = -0.1$ V?

4-12. In Problem 4-11, what is the value of V_o for each input if the strobe terminal is wired to ground via a 10-kΩ resistor?

4-13. Design a window detector circuit whose output is high when an input voltage lies between limits of $+6$ V and $+1$ V.

4-14. Which has a smaller propagation delay, a 301 op amp or a 311 comparator?

selected applications
of op amps

,ı₁,ıₗ,ıₗ,ıₗ,ₐ,ıₗ,ıₗ,ıₗ,ıₗ,ıₗ,ıₗ,ıₗ,ıₗ,ıₗ,ıₗ,ₐ,ıₗ,ıₗ,ıₗ,ıₗ,ıₗ,ıₗ,ıₗ,ıₗ,ıₗ,ıₗ,ıₗ,ₐ,ıₗ,ıₗ,ıₗᵘ

5

5-0 INTRODUCTION

Why is the op amp such a popular device? This chapter attempts to answer that question by presenting a wide selection of applications. They were selected to show that the op amp can perform as a very nearly ideal device. Moreover, the diversity of operations that the op amp can perform is almost without limit. In fact, applications that are normally very difficult, such as measuring short-circuit current, are rendered simple by the op amp. Together with a few resistors and a power supply, the op amp can, for example, measure the output from photodetectors, give audio tone control, equalize tones of different amplitudes, control high currents, and allow matching of semiconductor device charac-teristics, We begin with selecting an op amp circuit to make a high-resistance dc and ac voltmeter.

5-1 HIGH-RESISTANCE DC VOLTMETER

5-1.1 Basic Voltage-Measuring Circuit

Figure 5-1 shows a simple but very effective high-input-resistance dc volt-meter. The voltage to be measured, E_i, is applied to the $(+)$ input terminal. Since the differential input voltage is 0 V, E_i is developed across R_i. The meter current I_m is set by E_i and R_i just as in the noninverting amplifier.

$$I_m = \frac{E_i}{R_i} \tag{5-1}$$

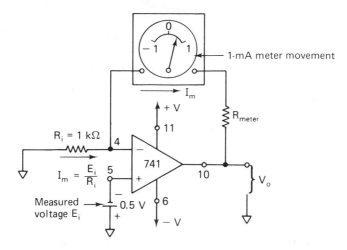

Figure 5-1   High-input-resistance dc voltmeter.

If R_i is 1 kΩ, then 1 mA of meter current will flow for $E_i = 1$ V dc. Therefore, the milliammeter can be calibrated directly in volts. As shown, this circuit can measure any dc voltage from -1 V to $+1$ V.

Example 5-1

Find I_m in Fig. 5-1.

Solution. From Eq. (5-1), $I_m = 0.5$ V/1 kΩ $= 0.5$ mA. The needle is deflected halfway between 0 and $+1$ mA.

One advantage of Fig. 5-1 is that E_i sees the very high input impedance of the $(+)$ input. Since the $(+)$ input draws negligible current, it will not load down or change the voltage being measured. Another advantage of placing the meter in the feedback loop is that if the meter resistance should vary, it will have no effect on meter current. Even if we added a resistor in series with the meter, within the feedback loop, it would not affect I_m. The reason is that I_m is set only by E_i and R_i. The output voltage will change if meter resistance changes, but in this circuit we are not concerned with V_o. This circuit is sometimes called a *voltage-to-current* converter.

5-1.2 Voltmeter Scale Changing

Since the input voltage in Fig. 5-1 must be less than the power supply voltages (± 15 V), a convenient maximum limit to impose on E_i is ± 10 V. The simplest way to convert Fig. 5-1 from a ± 1-V voltmeter to a ± 10-V voltmeter is to change R_i to 10 kΩ. In other words, pick R_i so that the full-scale input voltage E_{FS} equals R_i times the full-scale meter current I_{FS} or

$$R_i = \frac{E_{FS}}{I_{FS}} \qquad (5\text{-}2)$$

Example 5-2

A microammeter with 50 μA $= I_{FS}$ is to be used in Fig. 5-1. Calculate R_i
for $E_{FS} = 5$ V.
Solution. By Eq. (5-2), $R_i = 5$ V/50 μA $= 100$ kΩ. Before measuring higher
input voltages, use a voltage-divider circuit. The output of the divider is applied
to the (+) input.

5-2 UNIVERSAL HIGH-RESISTANCE VOLTMETER

5-2.1 Circuit Operation

The voltage-to-current converter of Fig. 5-2 can be used as a universal
voltmeter. That is, it can be used to measure positive or negative dc voltage or
the rms, peak, or peak-to-peak (p-p) value of a *sine wave*. To change from one
type of voltmeter to another, it is necessary to change only a single resistor. The
voltage to be measured, E_i, is applied to the op amp's (+) input. Therefore,
the meter circuit has a high input resistance.

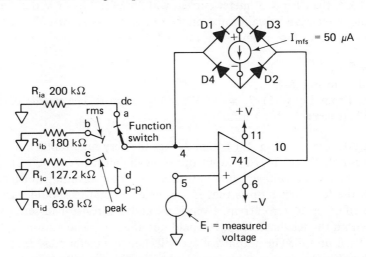

Figure 5-2 Basic high-resistance universal voltmeter circuit. The meaning
of a full-scale meter deflection depends on the function switch position as
follows: 10 V dc on position *a*, 10 V ac rms on position *b*, 10 V peak ac on
position *c*, and 10 V ac p-p on position *d*.

When E_i is positive, current flows through the meter movement and diodes
D_3 and D_4. When E_i is negative, current flows in the *same* direction through
the meter and diodes D_1 and D_2. Thus meter current direction is the same
whether E_i is positive or negative.

A dc meter movement measures the *average* value of current. Suppose that
a basic meter movement is rated to give full-scale deflection when conducting a
current of 50 μA. A voltmeter circuit containing the basic meter movement is

to indicate at full scale when E_i is a sine wave with a peak voltage of 10 V. The meter face should be calibrated linearly from 0 V to $+10$ V instead of 0 to 50 μA. The *circuit* and meter movement would then be called a *peak reading voltmeter* (for sine waves only) with a full-scale deflection for $E_{ip} = 10$ V. The following section shows how easy it is to design a universal voltmeter.

5-2.2 Design Procedure

The design procedure is as follows: Calculate R_i according to the application from one of the following equations:
(a) Dc voltmeter:

$$R_i = \frac{\text{full-scale } E_{dc}}{I_{mfs}} \tag{5-3a}$$

(b) Rms ac voltmeter (sine wave only):

$$R_i = 0.90 \frac{\text{full-scale } E_{rms}}{I_{mfs}} \tag{5-3b}$$

(c) Peak reading voltmeter (sine wave only)

$$R_i = 0.636 \frac{\text{full-scale } E_{peak}}{I_{mfs}} \tag{5-3c}$$

(d) Peak-to-peak ac voltmeter (sine wave only)

$$R_i = 0.318 \frac{\text{full-scale } E_{p/p}}{I_{mfs}} \tag{5-3d}$$

where I_{mfs} is the meter's full-scale current rating in amperes. The design procedure is illustrated by an example.

Example 5-3

A basic meter movement (such as the Simpson 260) is rated at 50 μA for full-scale deflection (with a meter resistance of 5 kΩ). Design a simple switching arrangement and select resistors to indicate full-scale deflection when the voltage to be measured is (a) 10 V dc; (b) 10 V rms; (c) 10 V peak; (d) 10 V p-p.
Solution. From Eqs. (5-3a) to (5-3d):
(a) $R_{ia} = \dfrac{10 \text{ V}}{50 \ \mu\text{A}} = 200 \text{ k}\Omega$ \qquad\qquad (b) $R_{ib} = 0.9 \dfrac{10 \text{ V}}{50 \ \mu\text{A}} = 180 \text{ k}\Omega$

(c) $R_{ic} = 0.636 \dfrac{10 \text{ V}}{50 \ \mu\text{A}} = 127.2 \text{ k}\Omega$ \qquad (d) $R_{id} = 0.318 \dfrac{10 \text{ V}}{50 \ \mu\text{A}} = 63.6 \text{ k}\Omega$

The resulting circuit is shown in Fig. 5-2.

It must be emphasized that neither meter resistance nor diode voltage drops affect meter current. Only R_i and E_i determine average or dc meter current.

5-3 VOLTAGE-TO-CURRENT CONVERTERS: FLOATING LOADS

5-3.1 Voltage Control of Load Current

From Sections 5-1 and 5-2 we learned not just how to make a voltmeter but that current in the feedback loop depends on the input voltage and R_i. There are applications where we need to pass a constant current through a load and hold it constant despite any changes in load resistance or load voltage. If the load does not have to be grounded, we simply place the load in the feedback loop and control both input and load current by the principle developed in Section 5-1.

5-3.2 Zener Diode Tester

Suppose that we have to test the breakdown voltage of a number of zener diodes at a current of precisely 5 mA. If we connect the zener in the feedback loop as in Fig. 5-3(a), our voltmeter circuit of Fig. 5-1 becomes a zener diode tester. That is, E_i and R_i set the load or zener current at a constant value. E_i forces V_o to go negative until the zener breaks down and clamps the zener voltage at V_z. R_i converts E_i to a current, and as long as R_i and E_i are constant, the load current will be constant regardless of the value of the zener voltage. Zener breakdown voltage can be calculated from V_o and E_i as $V_z = V_o - E_i$.

Example 5-4
In the circuit of Fig. 5-3(a), $V_o = 10.3$ V, $E_i = 5$ V, and $R_i = 1$ kΩ. Find (a) the zener current; (b) the zener voltage.
Solution. (a) From Eq. (3-1), $I = E_i/R_i$ or $I = 5$ V/1 kΩ $= 5$ mA. (b) From Fig. 5-3(a), rewrite the equation for V_o.

$$V_z = V_o - E_i = 10.3 \text{ V} - 5 \text{ V} = 5.3 \text{ V}$$

5-3.3 Diode Tester

Suppose that we needed to select diodes from a production batch and find pairs with matching voltage drops at a particular value of diode current. Place the diode in the feedback loop as shown in Fig. 5-3(b). E_i and R_i will set the value of I. The $(-)$ input draws negligible current, so I passes through the diode. As long as E_i and R_i are constant, current through the diode I will be constant

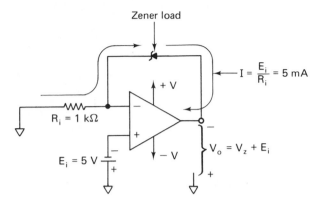

(a) Negligible current drawn from E_i, load current furnished by op amp

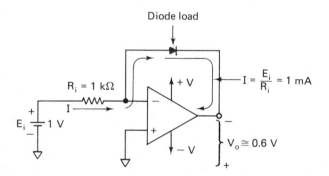

(b) Load current equals input current

Figure 5-3 Voltage-controlled load currents with loads in feedback loop.

at $I = E_i/R_i$. V_o will equal the diode voltage for the same reasons that V_o was equal to V_{R_f} in the inverting amplifier (see Section 3-1).

Example 5-5

$E_i = 1$ V, $R_i = 1$ kΩ, and $V_o = 0.6$ V in Fig. 5-3(b). Find (a) the diode current and (b) the voltage drop across the diode.

Solution. (a) $I = E_i/R_i = 1$ V/1 k$\Omega = 1$ mA. (b) $V_{diode} = V_o = 0.6$ V.

There is one disadvantage with the circuit of Fig. 5-3(b): E_i must be able to furnish the current. Both circuits in Fig. 5-3 can only furnish currents up to 10 mA because of the op amp's output current limitation. Higher-load currents can be furnished from the power supply terminal and a current boost transistor as shown in Fig. 5-4.

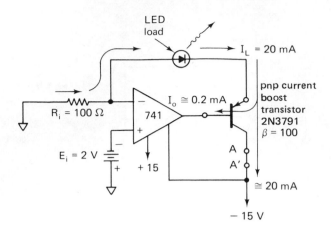

Figure 5-4   Voltage-to-high current converter.

5-4 LIGHT-EMITTING-DIODE TESTER

The circuit of Fig. 5-4 converts E_i to a 20-mA load current based on the same principles discussed in Sections 5-1 to 5-3. Since the 741's output terminal can only supply about 5 to 10 mA, we cannot use the circuits of Figs. 5-1 to 5-3 for higher load currents. But if we add a transistor as in Fig. 5-4, load current is furnished from the negative supply voltage. The op amp's output terminal is required to furnish only base current, which is typically $\frac{1}{100}$ of the load current. The factor $\frac{1}{100}$ comes from assuming that the transistor's beta equals 100. Since the op amp can furnish an output current of up to 5 mA into the transistor's base, this circuit can supply a maximum load current of 5 mA $\times$ 100 = 0.5 A.

A light-emitting diode such as the MLED50 is specified to have a typical brightness of 750 mcd provided that the forward diode current is 20 mA. E_i and R_i will set the diode current I_L equal to $E_i/R_i = 2$ V/100 Ω = 20 mA. Now brightness of LEDs can be measured easily one after another for test or matching purposes, because the current through each diode will be exactly 20 mA regardless of the LED's forward voltage.

It is worthwhile to note that a load of two LEDs can be connected in series with the feedback loop and both would conduct 20 mA. The load could also be connected in Fig. 5-4 between points AA' (which is in series with the transistor's collector) and still conduct about 20 mA. This is because the collector and emitter currents of a transistor are essentially equal. A load in the feedback loop is called a *floating load*. If one side of the load is grounded, it is a *grounded load*. To supply a constant current to a grounded load, another type of circuit must be selected, as shown in Section 5-5.

5-5 FURNISHING A CONSTANT CURRENT
TO A GROUNDED LOAD

5-5.1 Differential Voltage-to-Current Converter

The circuit of Fig. 5-5 can be called a differential voltage-to-current converter because the load current I_L depends on the *difference* between input voltage E_1 and E_2 and resistors R. I_L does *not* depend on load resistor R_L. Therefore, if E_1 and E_2 are constant, the grounded load is driven by a constant current. Load current can flow in either direction, so this circuit can either source or sink current.

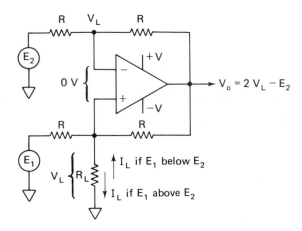

Figure 5-5 Differential voltage-to-current converter or constant-current source with grounded load.

Load current I_L is determined by

$$I_L = \frac{E_1 - E_2}{R} \qquad (5\text{-}4)$$

A positive value for I_L signifies that it flows downward in Fig. 5-5 and V_L is positive with respect to ground. A negative value of I_L means that V_L is negative with respect to ground and current flows upward.

Load voltage V_L (not I_L) depends on load resistor R_L from

$$V_L = I_L R_L \qquad (5\text{-}5)$$

To ensure that the op amp does not saturate, V_o must be known and can be calculated from

$$V_o = 2V_L - E_2 \qquad (5\text{-}6)$$

Circuit operation is illustrated by the following examples.

Example 5-6

In Fig. 5-5, $R = 10$ kΩ, $E_2 = 0$, $R_L = 5$ kΩ, and $E_1 = 5$ V. Find (a) I_L; (b) V_L; (c) V_o.

Solution. (a) From Eq. (5-4),

$$I_L = \frac{5\text{ V} - 0}{10\text{ k}\Omega} = 0.5\text{ mA}$$

(b) From Eq. (5-5),

$$V_L = 0.5\text{ mA} \times 5\text{ k}\Omega = 2.5\text{ V}$$

(c) From Eq. (5-6),

$$V_o = 2 \times 2.5\text{ V} = 5\text{ V}$$

Reversing the polarity of E_1 reverses I_L and the polarity of V_o and V_L.

Example 5-7

In Fig. 5-5, $R = 10$ kΩ, $E_2 = 5$ V, $R_L = 5$ kΩ, and $E_1 = 0$. Find (a)I_L; (b)V_L; (c)V_o. Compare this example with Example 5-6.

Solution. (a) From Eq.(5-4),

$$I_L = \frac{0 - 5\text{ V}}{10\text{ k}\Omega} = -0.5\text{ mA}$$

(b) From Eq. (5-5),

$$V_L = -0.5\text{ mA} \times 5\text{ k}\Omega = -2.5\text{ V}$$

(c) From Eq. (5-6),

$$V_o = 2(-2.5\text{ V}) - 5\text{ V} = -10\text{ V}$$

Note: V_L and I_L are reversed in polarity and direction, respectively, from Example 5-6. If the polarity of E_2 is reversed, I_L, V_L, and V_o change sign but *not* magnitude.

5-5.2 Constant-High-Current Source, Grounded Load

In certain applications, such as electroplating, it is desirable to furnish a high current, of constant value, to a grounded load. The circuit of Fig. 5-6 will furnish constant currents above 500 mA provided that the transistor is heat-sinked properly (above 5 W) and has a high beta ($\beta > 100$). The circuit operates as follows. The zener diode voltage is applied to one end of current sense resistor R_s and the op amp's positive input. Since the differential input voltage is 0 V, the zener voltage is developed across R_s. R_s and V_z set the emitter current, I_E, constant at V_z/R_s. The emitter and collector currents of a bipolar junction

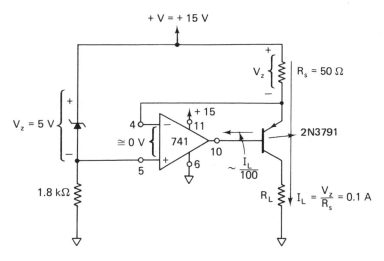

Figure 5-6   Constant-high-current source.

transistor are essentially equal. Since the collector current is load current I_L and $I_L \approx I_E$, the load current I_L is set by V_z and R_s.

If the op amp can furnish a base current drive of over 5 mA and if the beta of the transistor is greater than 100, then I_L can exceed 5 mA $\times$ 100 = 500 mA. The voltage across the load must not exceed the difference between the supply and the zener voltage; otherwise, the transistor and the op amp will go into saturation.

5-6   SHORT-CIRCUIT CURRENT MEASUREMENT AND CURRENT-TO-VOLTAGE CONVERSION

5-6.1   Introduction

Transducers such as phonograph pickups and solar cells convert some physical quantity into electrical signals. For convenience, the transducers may be modeled by a signal generator as in Fig. 5-7(a). It is often desirable to measure their maximum output current under short-circuit conditions; that is, we should place a short circuit across the output terminals and measure current through the short circuit. This technique is particularly suited to signal sources with very high internal resistance. For example, in Fig. 5-7(a), the short-circuit current I_{sc} should be 2.5 V/50 kΩ = 50 μA. However, if we place a microammeter across the output terminals of the generator, we no longer have a short circuit but a 5000-Ω resistance. The meter indication is

$$\frac{2.5 \text{ V}}{50 \text{ k}\Omega + 5 \text{ k}\Omega} \cong 45 \ \mu A$$

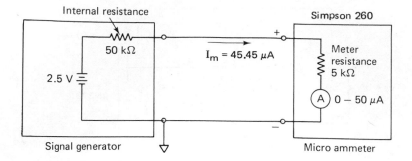

(a) Ammeter resistance reduces short-circuit current from the signal generator

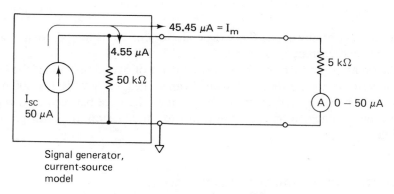

(b) Current-source model of signal generator in (a)

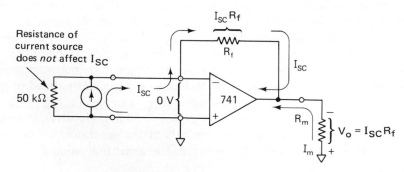

(c) Current-to-voltage converter

Figure 5-7 Current-measuring circuits.

High-resistance sources are better modeled by an equivalent Norton circuit. This model is simply the ideal short-circuit current, I_{SC}, in parallel with its own internal resistance as in Fig. 5-7(b). This figure shows how I_{SC} splits between its internal resistance and the meter resistance. To eliminate this current split, we will use the op amp.

5-6.2 Using the Op Amp to Measure Short-Circuit Current

The op amp circuit of Fig. 5-7(c) effectively places a short circuit around the current source. The $(-)$ input is at virtual ground because the differential input voltage is almost 0 V. The current source sees ground potential at both of its terminals, or the equivalent of a short circuit. *All* of I_{SC} flows toward the $(-)$ input and on through R_f. R_f converts I_{SC} to an output voltage, revealing the basic nature of this circuit to be a *current-to-voltage converter*.

Example 5-8

V_o measures 5 V in Fig. 5-7(c), and $R_f = 100\ \text{k}\Omega$. Find the short-circuit current I_{SC}.

Solution. From Fig. 5-7(c),

$$I_{SC} = \frac{V_o}{R_f} = \frac{5\ \text{V}}{100\ \text{k}\Omega} = 50\ \mu\text{A}$$

The resistance R_m is the resistance of either the voltmeter or the CRO. The current I_m needed to drive either instrument comes from the op amp and not from I_{SC}.

5-7 MEASURING CURRENT FROM PHOTODETECTORS

5-7.1 Photoconductive Cell

With the switch at position 1 in Fig. 5-8, a photoconductive cell, sometimes called a light-sensitive resistor (LSR), is connected in series with the $(-)$ input and E_i. The resistance of a photoconductive cell is very high in darkness and much lower when illuminated. Typically, its dark resistance is greater than 500 kΩ and its light resistance in bright sun is approximately 5 kΩ. If $E_i = 5$ V, then current through the photoconductive cell, I, would be 5 V/500 kΩ = 10 μA in darkness and 5 V/5 kΩ = 1 mA in sunlight.

Example 5-9

In Fig. 5-8 the switch is in position 1 and $R_f = 10\ \text{k}\Omega$. If the current through the photoconductive cell is 10 μA in darkness and 1 mA in sunlight, find V_o for (a) the dark condition; (b) the light condition.

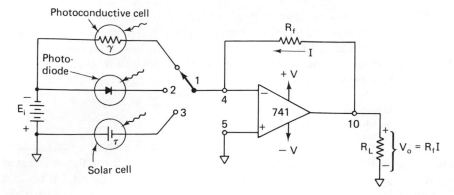

Figure 5-8   Using the op amp to measure output current from photodetectors.

Solution. From Fig. 5-8 $V_o = R_f I$. (a) $V_o = 10 \text{ k}\Omega \times 10 \text{ } \mu\text{A} = 0.1 \text{ V}$; (b) $V_o = 10 \text{ k}\Omega \times 1 \text{ mA} = 10 \text{ V}$. Thus the circuit of Fig. 5-8 converts the output current from the photoconductive cell into an output voltage (a current-to-voltage converter).

5-7.2 Photodiode

When the switch is in position 2 in Fig. 5-8, E_i is on one side of the photodiode and virtual ground on the other. The photodiode is reverse-biased, as it must be for normal operation. In darkness the photodiode conducts a small leakage current on the order of nanoamperes. But depending on the radiant energy striking the diode, it will conduct 50 μA or more. Therefore, current I depends only on the energy striking the photodiode and not on E_i. This current is converted to a voltage by R_f.

Example 5-10

With the switch in position 2 in Fig. 5-8 and $R_f = 100 \text{ k}\Omega$, find V_o as the light changes photodiode current from (a) 1 μA to (b) 50 μA.
Solution. From $V_o = R_f I_L$, (a) $V_o = 100 \text{ k}\Omega \times 1 \text{ } \mu\text{A} = 0.1 \text{ V}$; (b) $V_o = 100 \text{ k}\Omega \times 50 \text{ } \mu\text{A} = 5.0 \text{ V}$.

5-8 CURRENT AMPLIFIER

Characteristics of high-resistance signal sources were introduced in Section 5-6.1. There is no point in converting a current into an equal current, but a circuit that converts a current into a larger current can be very useful. The circuit of Fig. 5-9 is a current multiplier or current amplifier (technically, a current-to-current converter). The signal current source I_{sc} is effectively short-circuited by the input terminals of the op amps. All of I_{sc} flows through resistor mR, and the voltage across it is mRI_{sc}. (Resistor mR is known as a multiplying resistor and m the multiplier.) Since R and mR are in parallel, the voltage across

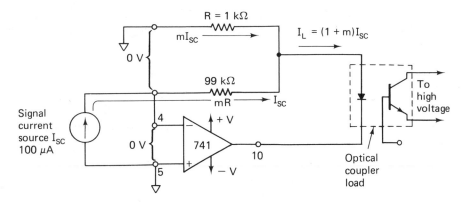

Figure 5-9   Current amplifier with optical coupler load.

R is also mRI_{sc}. Therefore, the current through R must be mI_{sc}. Both currents add to form load current I_L. I_L is an amplified version of I_{sc} and is found simply from

$$I_L = (1 + m)I_{sc} \qquad (5\text{-}7)$$

Example 5-10

In Fig. 5-9, $R = 1\ k\Omega$ and $mR = 99\ k\Omega$. Therefore, $m = 99k\Omega/1\ k\Omega = 99$. Find the current I_L through the emitting diode of the optical coupler.

Solution. By Eq. (5-7), $I_L = (1 + 99)(100\ \mu A) = 10$ mA.

It is important to note that the load does not determine load current. Only the multiplier m and I_{sc} determine load current. For variable current gain, mR and R can be replaced by a single 100-kΩ potentiometer. The wiper goes to the emitting diode, one end to ground and the other end to the $(-)$ input. The optical coupler isolates the op amp circuit from any high-voltage load.

5-9 SOLAR CELL ENERGY MEASUREMENTS

5-9.1 Introduction to the Problems

A solar cell (also called a photovoltaic cell) is a device that converts light energy directly into electrical energy. The best way to record the amount of energy received by the solar cell is to measure its short-circuit current. For example, one type of solar cell furnishes a short-circuit current I_{sc} that ranges from 0 to 0.5 A as sunlight varies from complete darkness to maximum brightness.

The problem facing users of these devices is to convert the 0 to 0.5-A solar cell output current to 0 to 10 V so that its performance can be monitored with a strip-chart recorder. Another problem is to measure 1/2 A of current with a low-current meter movement (0 to 0.1 mA). To solve this problem, I_{sc} must be

divided so that it can be measured on site with an inexpensive basic meter movement. The final problem is that the value of I_{sc} is too large to be used with the
op amp circuits studied thus far.

5-9.2   Converting Solar Cell Short-Circuit Current to a Voltage

The circuit of Fig. 5-10 solves several problems. First, the solar cell sees
the (−) input of the op amp as a virtual ground. Therefore, it can deliver its
short-circuit current I_{sc}. A second problem is solved when I_{sc} is converted by
R_f to a voltage V_o. To obtain a 0 to 10-V output for a 0 to 0.5-A input, R_f should
have a value of

$$R_f = \frac{V_o \text{ full scale}}{I_{sc} \text{ max}} = \frac{10 \text{ V}}{0.5 \text{ A}} = 20.0 \ \Omega$$

The solar cell current of 0.5 A is too large to be handled by the op amp. This
problem is solved by adding an *npn* current boost transistor.

The solar cell current flows through the emitter and collector of the boost
transistor to +V. Current gain of the transistor should exceed $\beta = 100$ to ensure
that the op amp has to furnish no more than 0.5 A/100 = 5 mA, when $I_{sc} =$
0.5 A.

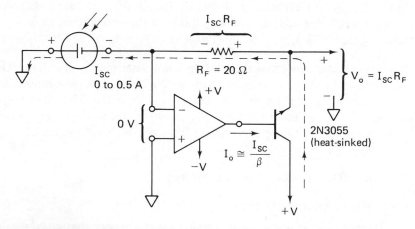

Figure 5-10   This circuit forces the solar cell to deliver a short-circuit current
I_{sc}. I_{sc} is converted to a voltage by R_f. Current boost is furnished by the *npn*
transistor.

5-9.3   Current-Divider Circuit (Current-to-Current Converter)

Only a slight addition to the circuit of Fig. 5-10 allows us to measure I_{sc}
with a low-current milliammeter or microammeter. The current-divider resistance dR_f is shown in Fig. 5-11. Resistance dR_f is made up of the meter resistance
R_m plus the scale resistor R_{scale}.

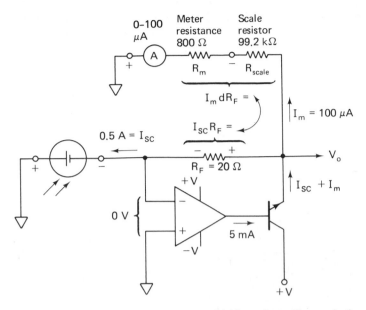

Figure 5-11 Current-to-current converter. Divider resistor dR_f equals the sum of meter resistance R_m and scale resistance R_{scale}. Short-circuit current $I_{sc} = 0.5$ A is converted by d down to 100 μA for measurement by a low-current meter.

The short-circuit current develops a voltage drop across R_f equal to V_o. V_o is also equal to the voltage across resistance dR_f. Thus the current divider d can be found by equating the voltage drops across dR_f and R_f.

$$V_o = I_{sc}R_f = I_m dR_f \qquad (5\text{-}8a)$$

so

$$d = \frac{I_{sc}}{I_m} \qquad (5\text{-}8b)$$

Example 5-11

If the meter in Fig. 5-11 is to indicate full scale at $I_m = 100$ μA when $I_{sc} = 0.5$ A, find resistance dR_f and R_{scale}.

Solution. From Eq. (5-8b), $d = 0.5$ A$/100$ μA $= 500$ and $dR_f = 500 \times 20\ \Omega = 100$ kΩ. Then $R_{scale} = dR_f - R_m = 100$ kΩ $- 0.8$ kΩ $= 99.2$ kΩ.

5-10 PHASE SHIFTER

5-10.1 Introduction

An ideal phase-shifting circuit should transmit a wave without changing its amplitude but changing its phase angle by a preset amount. For example, a sine wave E_i with a frequency of 1 kHz and peak value of 1 V is the input of the phase shifter in Fig. 5-12(a). The output V_o has the same frequency and amplitude but lags E_i by 90°. That is, V_o goes through 0 V 90° *after* E_i goes through 0 V. Mathematically, V_o can be expressed by $V_o = E_i \angle -90°$. A general expression for the output voltage of the phase-shifter circuit in Fig. 5-12(b) is given by

$$V_o = E_i \angle -\theta \qquad (5\text{-}9)$$

where θ is the phase angle and will be found from Eq. (5-10a).

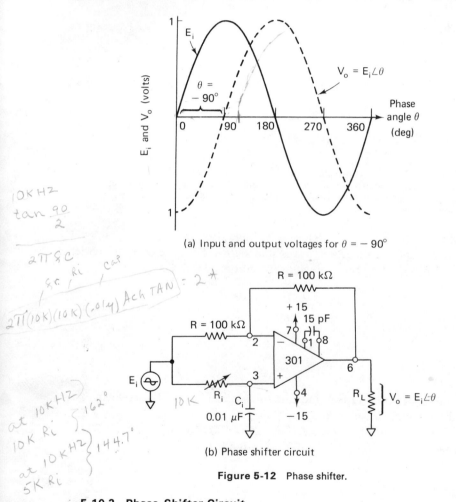

(a) Input and output voltages for $\theta = -90°$

(b) Phase shifter circuit

Figure 5-12   Phase shifter.

5-10.2   Phase-Shifter Circuit

One op amp, three resistors, and one capacitor are all that is required as shown in Fig. 5-12(b) to make an excellent phase shifter. The resistors R must be equal, and any convenient value from 10 to 220 kΩ may be used. Phase angle θ depends only on R_i, C_i, and the frequency f of E_i. The relationship is

$$\theta = 2 \arctan 2\pi f R_i C_i \qquad \text{(5-10a)}$$

where θ is in degrees, f in hertz, R_i in ohms, and C_i in farads. Equation (5-10a) is useful to find the phase angle if f, R_i, and C_i are known. If the desired phase angle is known, choose a value for C_i and solve for R_i:

$$R_i = \frac{\tan(\theta/2)}{2\pi f C_i} \qquad \text{(5-10b)}$$

Example 5-12

Find R_i in Fig. 5-12(b) so that V_o will lag E_i by 90°. The frequency of E_i is 1 kHz.

Solution. Since $\theta = -90°$, $\tan(-90°/2) = \tan(-45°) = -1$; from Eq. (5-10b),

$$R_i = \frac{1}{2\pi \times 1000 \times 0.01 \times 10^{-6}} = 15.9 \text{ k}\Omega$$

With $R_i = 15.9$ kΩ, V_o will have the phase angle shown in Fig. 5-12(a). This waveform is a negative cosine wave.

Example 5-13

If $R_i = 100$ kΩ in Fig. 5-12(b), find the phase angle θ.

Solution. From Eq. (5-9a),

$$\theta = 2 \arctan (2\pi)(1 \times 10^3)(100 \times 10^3)(0.01 \times 10^{-6})$$
$$= 2 \arctan 6.28$$
$$= 2 \times 81° = 162° \quad \text{and} \quad V_o = E_i \, \underline{/-162°}$$

It can be shown from Eq. (5-10a) that $\theta = -90°$ when R_i equals the reactance of C_i, or $1/(2\pi f C_i)$. As R_i is varied from 1 kΩ to 100 kΩ, θ varies from approximately $-12°$ to $-168°$. Thus, the phase shifter can shift phase angles over a range approaching 180°. If R_i and C_i are interchanged in Fig. 5-11(b), the phase angle is positive, and the circuit becomes a leading phase-angle shifter. The magnitude of θ is found from Eq. (5-10a), but the output is given by $V_o = E_i \underline{/\, 180° - \theta}$.

5-11 THE CONSTANT-VELOCITY RECORDING PROCESS

5-11.1 Introduction to Record-Cutting Problems

The process of recording data or music on a record is accomplished by a heated chisel-shaped cutting stylus that vibrates from side to side (laterally) in the record groove. Each groove is about 1 mil (0.001 in.) wide. The cutting stylus is vibrated by electromechanical transducers that are activated by the magnetic fields from a driving and a feedback coil. The coil–transducer–stylus combination is called the cutting head.

If the *amplitude* of the input signal current is held *constant*, the stylus cuts laterally at a constant velocity. When the frequency of the input signal is varied, the cutting rate or lateral *velocity* of the stylus remains *constant, provided that amplitude of the input signal is held constant.* This type of a recording process is called a *constant-velocity recording.*

5-11.2   Groove Modulation with Constant-Velocity Recording

Groove modulation is defined as the peak-to-peak lateral cutting distance and should depend only on the amplitude of the input signal, *not* on its frequency. Unfortunately, this is *not* the case for constant-velocity recording, as shown in Fig. 5-13.

If amplitude at the input signal is held constant at 10 mV, the cutter's lateral velocity V will be typically 2 in./s. The p-p lateral distance d can then be found from

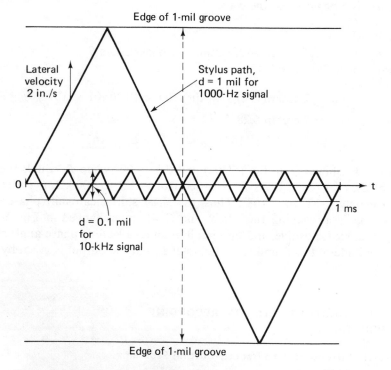

Figure 5-13   If signal amplitude is held constant across a cutting head, the resulting constant velocity recording process causes overcuts at low frequencies. The smaller cuts at high frequencies become indistinguishable from record noise due to surface imperfections.

$$d = \frac{V}{2f} \qquad (5\text{-}11)$$

where f is the signal frequency in hertz. The problem that results, because groove modulation depends on signal frequency, is brought out by an example.

Example 5-14

Find the peak-to-peak lateral cutting distance for signals of (a) 1000 Hz reference; (b) 10,000 Hz. Assume that the input signal is 10 mV and the stylus velocity is 2 in./s.

Solution   (a) From Eq. (5-11),

$$d = \frac{2 \text{ in.}}{s} \times \frac{1 \text{ (s)}}{(2)1000 \text{ (cycles)}} = 1 \text{mil}$$

(b) From Eq. (5-10),

$$d = \frac{2 \text{ in.}}{s} \times \frac{1 \text{ (s)}}{(2)10,000 \text{ (cycles)}} = 0.1 \text{ mil}$$

5-11.3   Record Cutover and Noise

Cutting paths for the stylus are shown in Fig. 5-13, for both 1-kHz and 10-kHz signals over a time period of 1 msec for a constant input voltage of 10 mV. There are two conclusions to be drawn from this figure. First, all frequencies below 1 kHz will cause cutover into the adjacent grooves. Second, the lateral cutting distance decreases as frequency increases, so that eventually the groove modulation will become indistinguishable from surface imperfections.

Suppose that the amplitude of the recording signal was increased to 100 mV, signifying a louder tone. This would cause a cutting velocity of 20 in./s. As shown in Fig. 5-14, this would result in cutting 10 grooves at 1 kHz. The problem with constant-velocity recording is summarized in Fig. 5-14. It shows the lack of dynamic range in fitting loud and soft tones of different frequencies into a 1-mil groove.

5-11.4   Solution to Record Cutover and Noise Problems

The solution to frequency cutover and noise problems is to attenuate the low-frequency signals and boost the high-frequency signals applied to the record cutter. The circuit that does this is called a *recording preequalizer*. For a given input signal, its output will cause the cutting stylus to cut a constant peak-to-peak distance independent of frequency. The process is shown conceptually in Fig. 5-15(b). With a preequalizer, the dynamic volume range can approach 40 dB or a distinction between loud and soft of 100:1.

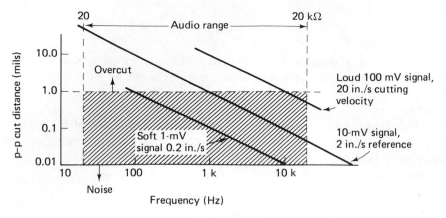

Figure 5-14 Acceptable constant-velocity recording process levels and frequencies lie within the crosshatched area.

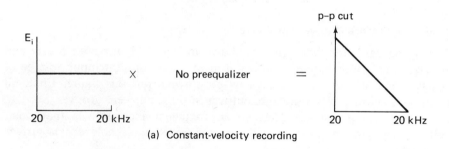

(a) Constant-velocity recording

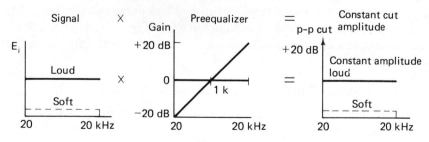

(b) Constant-velocity recording with preequalizer

Figure 5-15 When the signal to be recorded is transmitted through a pre-equalizer, the stylus gives a peak-to-peak cut that is independent of the signal frequency and depends only on the signal's amplitude.

5-12 RECORD PLAYBACK

5-12.1 Need for Playback Equalization

It was shown in Section 5-12 that constant-velocity cutting heads would make lateral cuts whose amplitude will depend on signal amplitude and not on signal frequency, *if* a *preequalizer* circuit was installed.

The stylus of a magnetic pickup cartridge moves a magnet within a coil. The output of the coil is proportional to the magnet's movement. The stylus will move faster (laterally) to follow groove modulation as frequency is increased. Therefore, output voltage of the pickup will increase directly with increasing frequency for the same peak-to-peak lateral cut (see Fig. 5-16).

Thus, if the output signal voltage at 1 kHz is just right, the output will get progressively lower at lower frequencies and progressively higher at higher frequencies. This means that the output of the magnetic cartridge must be equalized in amplitude by a *playback equalizer*. The circuit that does the amplitude equalization is called a *preamplifier* and its approximate relative and absolute gains are shown in Fig. 5-16. Low frequencies are amplified and high frequencies are attenuated. Thus the output of the equalizer gives a voltage that is proportional to amplitude of the lateral groove cut and not its frequency.

5-12.2 Preamplifier Gain and Signal Voltage Levels

The typical magnetic cartridge produces 5 mV of output for a needle velocity of 5 cm/s. The phonograph preamplifier must also provide different gains at different frequencies to bring this 5-mV output up to about 0.2 to 0.5 V at all frequencies to drive an audio amplifier. Typical *approximate* gains and signal levels are shown below for a velocity of 5 cm/s from a constant-amplitude cut at different frequencies.

Output of pickup (mV)	Gain of preamplifier (dB)		Output of preamplifier (V)	Frequency (Hz)
	Absolute	Relative		
0.05	1000	+20	0.5	20
5	100	0	0.5	1,000
50	10	−20	0.5	20,000

5-12.3 Playback Preamplifier Circuit Operation

The ideal RIAA (Record Industry Association of America) playback equalization curve is shown in Fig. 5-17(a) as a dashed line. An inexpensive circuit that accomplishes playback equalization is shown in Fig. 5-17(b). The RC4739 has two low-noise op amps on one chip. They are internally compen-

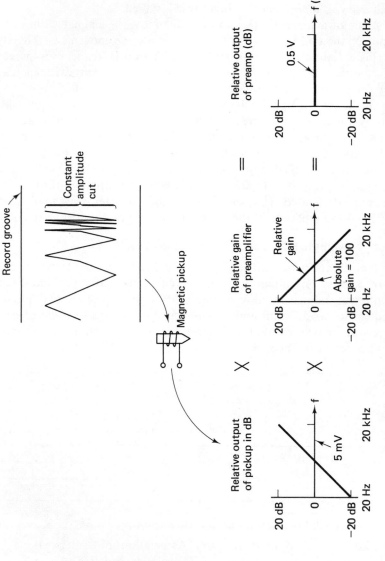

Figure 5-16 When a variable-frequency constant-amplitude groove modulation is applied to a magnetic pickup, its output voltage increases with increasing frequency. Gain of the playback equalizer decreases with increasing frequency so that its output is flat.

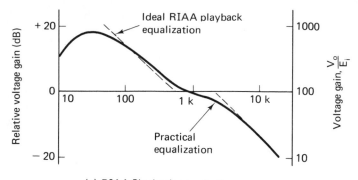

(a) RIAA Playback equalization curve

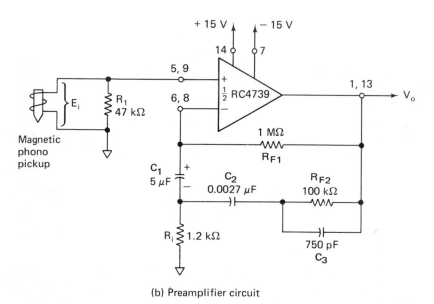

(b) Preamplifier circuit

Figure 5-17   RIAA playback equalization curve and preamplifier.

sated (see Chapter 10). (The μA 739 or MC 1303 may also be used as pin-for-pin replacements provided that external compensation is installed.) One RC4739 can equalize both channels of a stereo system. The first number on each terminal identifies the A-channel op amp and the second the B-channel op amp.

Operation of the circuit is analyzed by looking at the role of each capacitor:

1. At zero frequency (dc), all capacitors are open circuits and the gain $V_o/E_i = +1$.
2. As frequency is increased, the reactance of C_1 begins to decrease at 0.03 Hz and becomes negligible at about 26 Hz. In this low-frequency

range the gain increases from $+1$ to a value set by

$$A_{\text{CL}} = \frac{R_{\text{F1}} + R_{\text{i}}}{R_{\text{i}}} = 834$$

3. At 54 Hz the reactance of capacitor C_2 begins to decrease until at 580 Hz its reactance becomes negligible. (C_1 = short, C_3 = open in this frequency range.) R_{F1} is now connected in parallel with R_{F2} to reduce gain at 580 Hz to about 77.
4. When the frequency of E_{i} increases above 2.3 kHz, C_3 begins to bypass R_{F2} and R_{F1}, reducing gain at 20 dB/decade until the gain settles to unity at about 178 kHz.

The resultant practical playback equalization curve is shown in Fig. 5-17(a).

5-13 TONE CONTROL

5-13.1 Introduction

The preamplifier of Section 5-12 will deliver a flat frequency response at its output. In most high-fidelity systems, the owner wants to have a tone-control feature that allows boosting or cutting the volume of bass or treble frequencies. A frequency-controlling network, made of resistors and capacitors, could be installed in series with the output of the preamplifier. However, this network would attenuate some of the frequencies by as much as 1/100 or 20 dB. Much of the gain so carefully built into the preamplifier would be lost.

5-13.2 Tone-Control Circuit

The practical tone-control circuit shown in Fig. 5-18(a) (1) features boost or cut of bass frequencies below 500 Hz and of treble frequencies above 2 kHz and (2) eliminates attenuation. The top 50-kΩ audio taper potentiometer is the bass frequency control. With the wiper adjusted to full boost position, the voltage gain at 10 Hz is about $10R/R$ or 10. With the wiper at full bass cut, the voltage gain at 10 Hz is about $R/10R = 0.1$. In effect, the $10R$ pot is adjusted to be in series with R_{i} for cut or R_{f} for boost. The boost capacitors C_B begin to bypass the pot at frequencies between 50 Hz and 500 Hz, as shown in Fig. 5-18(b).

When adjusted to full boost, the treble control, $R/3$, and the capacitors C_T set the gain at 20 kHz to 10. At full cut, the gain is 0.1, as shown in Fig. 5-18(b). With both bass and treble control pots adjusted to the center of their rotation, the frequency response of the tone-control circuit will be flat. The input signal E_{i} delivered from the preamplifier should be about 0.2 V rms at 1 kHz. Therefore, the output of the tone control should be at about the same level.

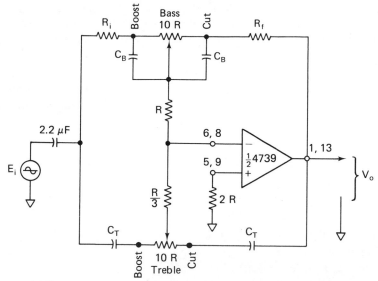

(a) Tone-control circuit; $C_B = C_T = 0.068 \ \mu F$, $R_i = R_f = R = 5 \ k\Omega$.
Connections for $+V$ and $-V$ shown in Figure 5.12b.

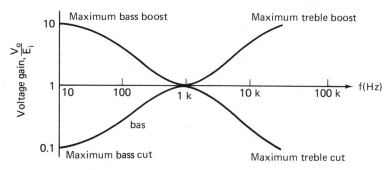

(b) Tone-control circuit frequency response curves

Figure 5-18 The tone-control circuit in (a) has the frequency-response curves shown in (b).

More applications for the 741 and 301 are shown in Appendices 1 and 2, respectively.

PROBLEMS

5-1. In the circuit of Fig. 5-1, if $E_i = 0.75$ V, find I_m.

5-2. Repeat Example 5-2 for $E_{FS} = 10$ V.

5-3. A microammeter with $I_{FS} = 100 \ \mu A$ is used in Fig. 5-1. If $R_i = 40 \ k\Omega$, find E_{FS}.

5-4. In the basic high-resistance universal voltmeter of Fig. 5-2, $I_{mfs} = 100\ \mu A$. Calculate (a) R_{ia} for a full-scale dc voltage of 10 V; (b) R_{ib} for an rms ac full-scale voltage of 10 V; (c) R_{ic} for a peak ac voltage of 10 V; (d) R_{id} for a p-p ac voltage of 10 V.

5-5. In Fig. 5-2, which diodes conduct when (a) E_i is positive; (b) E_i is negative?

5-6. Is Fig. 5-2 classified as a voltage-to-voltage or a voltage-to-current converter?

5-7. If the diode current is to be 5 mA in Fig. 5-3(b), you could either (a) change E_i to _____ or (b) change R_i to _____.

5-8. If $V_o = 11$ V and $E_i = 5$ V in Fig. 5-3, find V_z.

5-9. I_1 must equal 20 mA in Fig. 5-4 when $E_i = -10$ V. Find R_i.

5-10. Define a floating load.

5-11. In Fig. 5-5, $E_2 = 0$ V, $R_L = 1$ kΩ and $R = 10$ kΩ. Find I_L, V_L, and V_o for (a) $E_1 = -2$ V; (b) $E_1 = +2$ V.

5-12. In Fig. 5-5, $E_1 = 0$ V. $R_L = 1$ kΩ and $R = 10$ kΩ. Find I_L, V_L, and V_o for (a) $E_2 = -2$ V; (b) $E_2 = +2$ V.

5-13. In Fig. 5-5, $E_1 = E_2 = -5$ V. Find I_L, V_L, and V_o.

5-14. Replace V_z in Fig. 5-6 with a 1.8-kΩ resistor. Find I_L.

5-15. Sketch an op amp circuit that will draw the short-circuit current from a signal source and convert the short-circuit current to a voltage.

5-16. A CL5M9M photocell has a resistance of about 10 kΩ under an illumination of 2 fc. If $E_i = -10$ V in Fig. 5-8, calculate R_F for an output V_o of 0.2 V when the photoconductive cell is illuminated by 2 fc.

5-17. Change multiplier resistor mR in Fig. 5-9 to 49 kΩ. Find I_L.

5-18. A solar cell is installed in the circuit of Fig. 5-11 that has a maximum short-circuit current of 0.1 A $= I_{SC}$. (a) Select R_F to give $V_o = 10$ V when $I_{SC} = 0.1$ A. (b) A 50-μA meter movement is to indicate full scale when $I_{SC} = 0.1$ A. Find R_{scale} if $R_M = 5$ kΩ.

5-19. Resistor R_i is changed to 10 kΩ in Example 5-13. Find the phase angle θ.

5-20. Find the peak-to-peak lateral cutting distance in Example 5-14 at a signal frequency of 20 kHz.

5-21. Does a record playback preamplifier circuit provide greater amplification for the lower frequencies or the higher frequencies?

5-22. In the tone-control circuit of Fig. 5-18, what is the (a) gain at 10 Hz when the base control is at full boost; (b) gain at 20 kHz when the treble control is at full cut?

signal generators

6

6-0 INTRODUCTION

Up to now our main concern has been to use the op amp in circuits that process signals. In this chapter we concentrate on op amp circuits that generate signals. Four of the most common and useful signals are described by their shape when viewed on a cathode ray oscilloscope. They are the square wave, triangular wave, sawtooth wave, and sine wave. Accordingly, the signal generator is classified by the shape of the wave it generates. Some circuits are so widely used that they have been assigned a special name. For example, the first circuit presented in Section 6-1 is a multivibrator that generates primarily square waves.

6-1 FREE-RUNNING MULTIVIBRATOR

6-1.1 Multivibrator Action

A *free-running* or *astable multivibrator* is a square-wave generator. The circuit of Fig. 6-1 is a multivibrator circuit and looks something like a comparator with hysteresis (Chapter 4), except that the input voltage is replaced by a capacitor. Resistors R_1 and R_2 form a voltage divider to feed back a fraction of the output to the (+) input. When V_o is at $+V_{sat}$, as shown in Fig. 6-1(a), the feedback voltage is called the upper-threshold voltage V_{UT}. V_{UT} is given in Eq. (4-1) and repeated here for convenience:

$$V_{UT} = \frac{R_2}{R_1 + R_2}(+V_{sat}) \qquad (6\text{-}1)$$

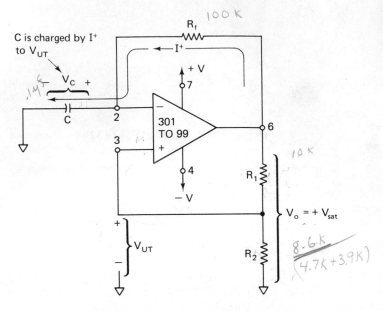

(a) When C is charged to V_{UT}, V_o switches to $- V_{sat}$

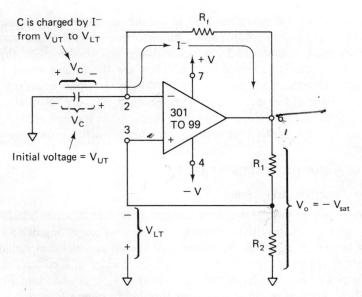

(b) When C is charged to V_{LT}, V_o switches to $+ V_{sat}$

Figure 6-1 Free-running multivibrator ($R_1 = 100$ kΩ, $R_2 = 86$ kΩ). Output-voltage waveform shown in Fig. 6-2.

Resistor R_f provides a feedback path to the $(-)$ input. When V_o is at $+V_{sat}$, current I^+ flows through R_f to charge capacitor C. As long as the capacitor voltage V_C is less than V_{UT}, the output voltage remains at $+V_{sat}$.

When V_C charges to a value slightly greater than V_{UT}, the $(-)$ input goes positive with respect to the $(+)$ input. This switches the output from $+V_{sat}$ to $-V_{sat}$. The $(+)$ input is now held negative with respect to ground because the feedback voltage is negative and given by

$$V_{LT} = \frac{R_2}{R_1 + R_2}(-V_{sat}) \qquad (6\text{-}2)$$

Equation (6-2) is the same as Eq. (4-2). Just after V_o switches to $-V_{sat}$, the capacitor has an initial voltage equal to V_{UT} [see Fig. 6-1(b)]. Now current I^- discharges C to 0 V and recharges C to V_{LT}. When V_C becomes slightly more negative than the feedback voltage V_{LT}, output voltage V_o switches back to $+V_{sat}$. The condition in Fig. 6-1(a) is reestablished except that C now has an initial charge equal to V_{LT}. The capacitor will discharge from V_{LT} to 0 V and then recharge to V_{UT}, and the process is repeating. Free-running multivibrator action is summarized as follows:

1. When $V_o = +V_{sat}$, C charges from V_{LT} to V_{UT} and switches V_o to $-V_{sat}$.
2. When $V_o = -V_{sat}$, C charges from V_{UT} to V_{LT} and switches V_o to $+V_{sat}$.

The time needed for C to charge and discharge determines the frequency of the multivibrator.

6-1.2 Frequency of Oscillation

The capacitor and output voltage waveforms for the free-running multivibrator are shown in Fig. 6-2. Resistor R_2 is chosen to equal $0.86R_1$ to simplify calculation of capacitor charge time. Time intervals t_1 and t_2 show how V_C and V_o change with time for Fig. 6-1(a) and (b), respectively. Time intervals t_1 and t_2 are equal to the product of R_f and C.

The period of oscillation, T, is the time needed for one complete cycle. Since T is the sum of t_1 and t_2,

$$T = 2R_fC \quad \text{for } R_2 = 0.86R_1 \qquad (6\text{-}3a)$$

The frequency of oscillation f is the reciprocal of the period T and is expressed by

$$f = \frac{1}{T} = \frac{1}{2R_fC} \qquad (6\text{-}3b)$$

where T is in seconds, f in hertz, R_f in ohms, and C in farads.

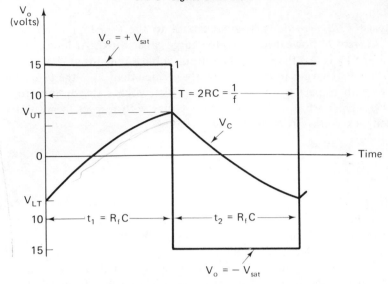

Figure 6-2   Voltage waveshapes for the multivibrator of Fig. 6-1.

Example 6-1

In Fig. 6-1, if $R_1 = 100$ kΩ, $R_2 = 86$ kΩ, $+V_{\text{sat}} = +15$ V, and $-V_{\text{sat}} = -15$ V, find (a) V_{UT}; (b) V_{LT}.

Solution. (a) By Eq. (6-1),

$$V_{\text{UT}} = \frac{86 \text{ k}\Omega}{186 \text{ k}\Omega} \times 15 \text{ V} \approx 7 \text{ V}$$

(b) By Eq. (6-2),

$$V_{\text{LT}} = \frac{86 \text{ k}\Omega}{186 \text{ k}\Omega}(-15 \text{ V}) = -7 \text{ V}$$

Example 6-2

Find the period of the multivibrator in Example 6-1 if $R_f = 100$ kΩ and $C = 0.1$ μF.

Solution. Using Eq. (6-3a), $T = (2)(100 \text{ k}\Omega)(0.1 \ \mu\text{F}) = 0.020 \text{ s} = 20 \text{ ms}$.

Example 6-3

Find the frequency of oscillation for the multivibrator of Example 6-2.

Solution. From Eq. (6-3b),

$$f = \frac{1}{20 \times 10^{-3} \text{ s}} = 50 \text{ Hz}$$

Example 6-4

Show why $T = 2R_f C$ when $R_2 = 0.86R$, as stated in Eq. (6-3a).

Solution. The time t required for a capacitor C to charge through a resistor R_F from some *starting* capacitor voltage toward some *aiming* voltage to a *stop* voltage is expressed generally as

$$t = R_f C \ln \left(\frac{\text{aim} - \text{start}}{\text{aim} - \text{stop}} \right)$$

Applying the equation to Fig. 6-2 yields

$$t_1 = R_f C \ln \left(\frac{+V_{sat} - V_{LT}}{+V_{sat} - V_{UT}} \right)$$

If the magnitudes of $+V_{sat}$ and $-V_{sat}$ are equal, the term in parentheses simplifies to

$$\ln \left[\frac{+V_{sat} - \dfrac{R_2}{R_1 + R_2}(-V_{sat})}{+V_{sat} - \dfrac{R_2}{R_1 + R_2}(+V_{sat})} \right] = \ln \left(\frac{R_1 + 2R_2}{R_1} \right)$$

Since the ln 2.778 $= 1$, the ln term can be reduced to 1 if

$$\frac{R_1 + 2R_2}{R_1} = 2.718 \qquad \text{or} \qquad R_2 = 0.86R_1$$

Now $t_1 = R_f C$ and $t_2 = R_f C$ if $R_2 = 0.86R_1$. Therefore, $T = t_1 + t_2 = 2R_f C$.

6-2 ONE-SHOT MULTIVIBRATOR

6-2.1 Introduction

A *one-shot multivibrator* generates a single output pulse in response to an input signal. The length of the output pulse depends only on external components (resistors and capacitors) connected to the op amp. As shown in Fig. 6-3, the one-shot generates a single output pulse on the negative-going edge of E_i. The duration of the input pulse can be longer or shorter than the expected output pulse. The duration of the output pulse is represented by τ in Fig. 6-3. Since τ can be changed only by changing resistors or capacitors, the one-shot can be considered a *pulse stretcher*. This is because the width of the output pulse is wider than the input pulse. Moreover, the one-shot introduces an idea of an adjustable delay, that is, the delay between the time when E_i goes negative and the time for V_o to go positive again. Operation of the one-shot will be studied in three parts: (1) the stable state, (2) transition to the timing state, and (3) the timing state.

6-2.2 Stable State

In Fig. 6-4(a), V_o is at $+V_{sat}$. Voltage divider R_1 and R_2 feeds back V_{UT} to the (+) input. V_{UT} is given by Eq. (6-1). The diode D_1 clamps the (−) input at approximately $+0.5$ V. The (+) input is positive with respect to the (−)

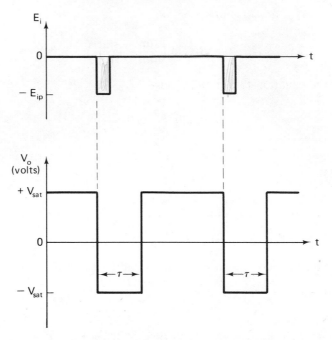

Figure 6-3 Input signal E_i and output pulse of a one-shot multivibrator.

input, and the high open-loop gain times the differential input voltage ($E_d =$ 2.1 − 0.5 = 1.6 V) holds V_o at $+V_{sat}$.

6-2.3 Transition to the Timing State

If input signal E_i is at a steady dc potential as in Fig. 6-4(a), the (+) input remains positive with respect to (−) input and V_o stays at $+V_{sat}$. However, if E_i goes negative by a peak value E_{ip} approximately equal to twice V_{UT}, the voltage at the (+) input will be pulled below the voltage at the (−) input. Once the (+) input becomes negative with respect to the (−) input, V_o switches to $-V_{sat}$. With this change, the one-shot is now in its timing state. For best results, C_i should be greater than 0.005 μF.

6-2.4 Timing State

The timing state is an unstable state; that is, the one-shot cannot remain very long in this state for the following reasons. Resistors R_1 and R_2 in Fig. 6-4(b) feed back a negative voltage ($V_{LT} = -2.1$ V) to the (+) input. The diode D_1 is now reversed-biased by $-V_{sat}$ and is essentially an open circuit. Capacitor C discharges to 0 and then recharges with a polarity opposite to that in Fig. 6-4(a) [see Fig. 6-4(b)]. As C recharges, the (−) input becomes more and more negative with respect to ground. When the capacitor voltage is slightly more negative than V_{LT}, V_o switches to $+V_{sat}$. The one-shot has now completed

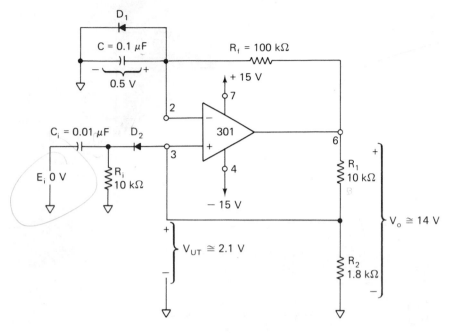

(a) Stable state of a one-shot multivibrator

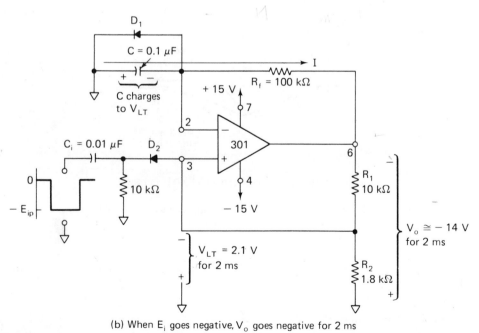

(b) When E_i goes negative, V_o goes negative for 2 ms

Figure 6-4 Monostable or one-shot multibibrator.

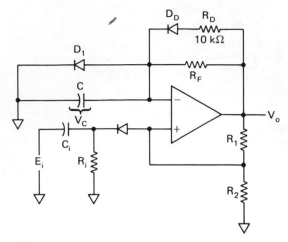

(a) Discharge diode D_D and R_D are added to Fig. 6-4

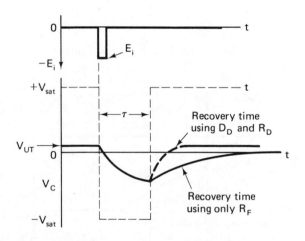

(b) Recovery time is reduced by D_D and R_D

Figure 6-5 The recovery time of a one-shot multivibrator is reduced by adding discharge diode D_D and R_D. R_D should be about one-tenth of R_F to reduce recovery time by one-tenth.

its output pulse and is back to the stable state in Fig. 6-4(a). Since the one-shot has only one stable state, it is also called a *monostable multivibrator*.

6-2.5 Duration of Output Pulse

If R_2 is made about one-fifth of R_1, in Fig. 6-4, then the duration of output pulse is given by

$$\tau \approx \frac{R_f C}{5} \tag{6-4}$$

Example 6-5

Calculate τ for the one-shot of Fig. 6-4.

Solution. By Eq. (6-4),

$$\tau = \frac{(100 \text{ k}\Omega)(0.1 \text{ } \mu\text{F})}{5} = 2 \text{ ms}$$

For test purposes, E_i can be obtained from a square-wave or pulse genera-tor. Diode D_2 prevents the one-shot from coming out of the timing state on positive transitions of E_i. To build a one-shot that has a positive output pulse for a positive input signal, simply reverse the diodes.

6-2.6 Recovery Time

After the timing state is completed, the output returns to $+V_{\text{sat}}$. However, the circuit is not ready to be retriggered reliably until C returns to its initial state of 0.5 V because it takes time for C to be discharged from $V_{\text{LT}} = -2.1$ V in Fig. 6-4(b) to 0.5 V in Fig. 6-4(a). This time interval is called *recovery time* and is shown in Fig. 6-5(b). Recovery time is approximately τ.

Normally, C is charged back to its initial state by a current through R_F. By adding a discharge resistor R_D in parallel with R_F, as in Fig. 6-5(a), the recovery time is reduced. Typically, if $R_D = 0.1R_F$, recovery time is reduced by one-tenth. Diode D_3 eliminates R_D from affecting the timing-cycle interval τ.

6-3 GENERATING TRIANGLE AND SAWTOOTH WAVES WITH A MULTIVIBRATOR

6-3.1 Introduction

The signal voltage waveshapes of the free-running multivibrator of Fig. 6-1 can be modified by changing components in the negative feedback loop. Three typical component arrangements have been selected to illustrate the versatility of the basic multivibrator.

6-3.2 Triangle-Wave Generator

A bridge rectifier and field-effect transistor (FET) are connected in the negative feedback loop of Fig. 6-6. The drain terminal D of the FET is always positive with respect to its source S. A resistor R_F is connected in series with the source and gate terminals so that the FET functions as a *constant-current source*. The magnitude of this current source I is determined by the value of R_F and characteristics of the FET, provided that the drain–source voltage

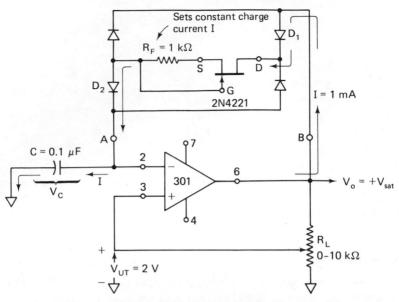

(a) When $V_o = +V_{sat}$, V_C charges linearly from V_{LT} to V_{UT} and switches V_o to $-V_{sat}$

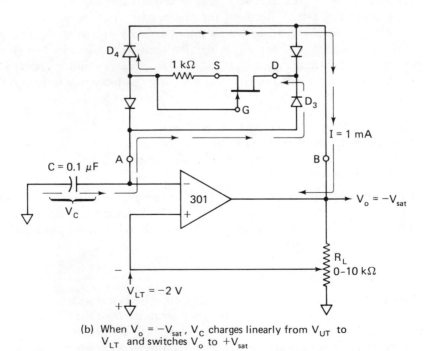

(b) When $V_o = -V_{sat}$, V_C charges linearly from V_{UT} to V_{LT} and switches V_o to $+V_{sat}$

Figure 6-6 A diode bridge and FET convert a multivibrator to a triangle-wave generator.

exceeds about 3 V. Typical values of I for a 2N4221 FET are as follows:

R_F	0 Ω	0.5 kΩ	1 kΩ	5 kΩ	10 kΩ
I	3.2 mA	1.4 mA	1 mA	0.3 mA	0.16 mA

Circuit operation is analyzed by assuming that V_o is $+V_{sat}$, as in Fig. 6-6(a). Diodes D_1 and D_2 conduct constant current I to charge capacitor C linearly from V_{LT} to V_{UT}. See time interval t_p in Fig. 6-7. V_{UT} and V_{LT} are set by the potentiometer R_L. When capacitor voltage V_C equals V_{UT}, V_o switches to $-V_{sat}$ as shown in Fig. 6-6(b). V_C then charges linearly from V_{UT} down to V_{LT} and switches V_o back to $+V_{sat}$. See time t_n in Fig. 6-7. This completes one period of oscillation and the sequence repeats to generate the waveshapes shown in Fig. 6-7.

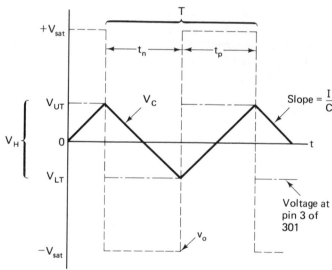

Figure 6-7 Waveshapes for the triangular-wave generator of Fig. 6-6. Capacitor voltage V_C is a triangular wave. Two square-wave outputs V_0 and V_{01} are also available.

The frequency of oscillation is designed or analyzed by the equation

$$f = \frac{I}{2CV_H} = \frac{1}{T} \tag{6-5}$$

where units are f in hertz, I in amperes, C in farads, and hysteresis voltage V_H in volts.

Example 6-6

V_{UT} and V_{LT} are adjusted to $+2$ V and -2 V, respectively, in Fig. 6-6. If $I = 1$ mA and $C = 0.1$ μF, find the frequency of oscillation. Note that $V_H = V_{UT} - V_{LT} = 4$ V.

Solution. From Eq. (6-5),

$$f = \frac{1 \times 10^{-3} \text{ A}}{2(0.1 \times 10^{-6} \text{ F})(4 \text{ V})} = 1250 \text{ Hz}$$

Observe from both Figs. 6-6 and 6-7 that an adjustment of potentiometer R_L changes both the p-p amplitude of the triangular wave (V_H) *and* the frequency. If R_L is held constant, the frequency can be adjusted by replacing R_F with a variable resistor to adjust I. Doubling I will double the frequency.

6-4 TRIANGULAR-WAVE GENERATOR WITH INDEPENDENTLY ADJUSTABLE SLOPES

Two *FET current generators* and two diodes are shown in Fig. 6-8(a). They can replace the negative feedback network in Fig. 6-6 to make a triangular-wave generator that has independently adjustable slopes. As shown in Fig. 6-8(a), when $V_o = +V_{sat}$, R_p is used to adjust I_p and capacitor C charges from V_{LT} to V_{UT} in a time t_p determined by

$$t_p = \frac{CV_H}{I_p} \tag{6-6a}$$

When V_C reaches V_{UT}, V_o switches to $-V_{sat}$ and V_C charges from V_{UT} to V_{LT} in time t_n as determined by

$$t_n = \frac{CV_H}{I_n} \tag{6-6b}$$

When V_C reaches V_{LT}, V_o switches to $-V_{sat}$ and the sequence repeats to generate a frequency whose period T is found from

$$T = t_p + t_n \tag{6-6c}$$

6-5 SAWTOOTH-WAVE GENERATOR

A short circuit is substituted for the negative adjustable constant current generator $R_N - Q_N$ in Fig. 6-8(a) to make a *positive-going sawtooth wave generator* in Fig. 6-8(b). Q_p, R_p, and D_p set the positive rise time t_p as given by

$$t_p = \frac{C(V_{UT} + 0.6 \text{ V})}{I_p} \tag{6.7}$$

When V_C rises to V_{UT}, V_o goes negative, allowing diode D_N to rapidly discharge capacitor C from V_{UT} to 0.6 V. The discharge occurs in a time interval set by the propagation delay of the 301 (Chapter 4) and its slew rate (Chapter 10), and is typically equal to 15 μs.

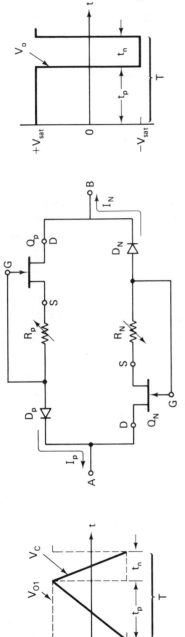

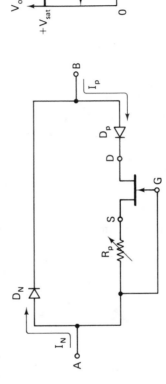

(a) Installing this circuit in the negative-feedback loop of Fig. 6-6 allows independent adjustment of positive rise time t_p by R_p and negative fall time t_n by R_N

(b) By substituting a short circuit for Q_N and R_N in (a), the circuit operates as a sawtooth-wave generator

Figure 6-8 By changing components in the negative-feedback loop of Fig. 6-6 (terminals A and B) the circuit operates as a triangular wave with independently adjustable slopes in (a) or as a sawtooth-wave generator in (b).

If the Q_p constant-current generator was replaced by a short circuit in Fig. 6-8(a) instead of Q_N, the result would be a *negative-going sawtooth-wave generator*.

6-6 TIMERS AND TRIANGLE-WAVE GENERATORS WITH THE INTEGRATOR

6-6.1 The Integrator

The free-running multivibrators of the preceding sections were characterized by a capacitor in the op amp's input circuit. If a capacitor is connected in the feedback loop, the circuit is classified as an *integrator* for reasons that will be explained in Section 6-6.3. The advantage of integrator type circuits is that the feedback capacitor is charged by a constant current that can be controlled easily by a grounded voltage source. Before studying applications of integrators, we examine how a constant charging current controls capacitor voltage and also how the capacitor voltage can be used to indicate elapsed time.

6-6.2 Ramp-Generator Theory

The constant-current source in Fig. 6-9 generates a voltage across the capacitor V_C that increases directly with elapsed time. The waveshape of V_C versus time is descriptively called a *ramp*. The capacitor charge Q depends on the constant current I and elapsed time t in accordance with

$$Q = It \qquad (6\text{-}8a)$$

An equation relating voltage with charge and capacitance is

$$Q = CV_C \qquad (6\text{-}8b)$$

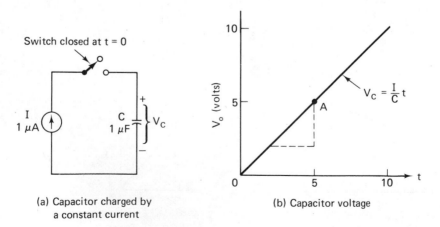

(a) Capacitor charged by a constant current

(b) Capacitor voltage

Figure 6-9 Voltage across a capacitor rises at a constant rate when the capacitor is charged by a constant current.

Substituting Eq. (6-8a) into Eq. (6-8b) to eliminate Q yields

$$V_C = \frac{I}{C} \times t \qquad (6\text{-}8c)$$

where V_C is in volts, t in seconds, I in amperes, and C in farads. If we know I and C, the capacitor voltage V_C will be directly related to time elapsed after the switch is closed.

Example 6-7

Calculate the capacitor voltage V_C in Fig. 6-9(a) 5 s after the switch is closed.
Solution. By Eq. (6-8c),

$$V_C = \frac{1 \ \mu A}{1 \ \mu F}(5 \text{ s}) = 5 \text{ V}$$

Figure 6-9b is a general plot of V_C versus time. The solution to Example 6-7 is point A on this diagram. Equation (6-8c) may be rearranged to show the *rate* at which V_o rises or falls in volts per second.

$$\frac{V_C}{t} = \frac{I}{C} = \frac{1 \ \mu A}{1 \ \mu F} = 1 \ \frac{V}{s} \qquad (6\text{-}8d)$$

V_C represents a continuous tally of how much charge has been stored by the capacitor. For example, after the first second, $V_C = 1$ V and the capacitor has stored 1 microcoulomb (μC) of charge. For each subsequent second the capacitor adds another microcoulomb. So V_C represents the sum of accumulated charge over a period of time. In mathematics this type of summing process is called *integration*. Hence this circuit is called an *integrator*. The triangular shape of V_C is called a *ramp* and is the basis for generating many useful control signals. To make a basic single-ramp generator, we will use the op amp.

6-6.3   Ramp-Generator Circuit

The current source in Fig. 6-9(a) is replaced by input voltage E_i and R_i and an op amp as shown in Fig. 6-10. Current I is set by E_i and R_i at $I = E_i/R_i$. Substituting for I in Eq. (6-8c), we obtain V_o in terms of E_i and time t:

$$V_o = -E_i \times \frac{1}{R_i C} \times t \qquad (6\text{-}9)$$

where R_i is in ohms, C in farads, t in seconds, and V_o and E_i in volts. The minus sign in Eq. (6-9) shows that E_i is applied through R_i to the $(-)$ input. Note that the capacitor voltage V_C equals V_o, so now the load current will be furnished from the op amp's output terminal and will not drain the capacitor.

There are two disadvantages to the circuit of Fig. 6-10. The first and most obvious is that V_o can go negative only to $-V_{sat}$. The second, not so obvious,

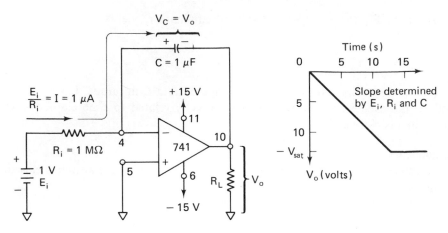

Figure 6-10   Single-ramp generator, a basic integrator.

is the fact that V_o will not remain at 0 V when $E_i = 0$ V. The reason for this is the inevitable presence of small bias currents that will charge the capacitor. (Bias currents will be studied in Chapter 9.) One method of stopping the capacitor from charging is to place a short circuit across it. V_C and V_o will then stay at 0 V. To start the ramp again, simply remove the short.

If a positive-going ramp is needed, simply reverse E_i. Understanding how a single-ramp generator works is needed in designing adjustable timing circuits.

6-7 ADJUSTABLE TIMER

6-7.1 Circuit Description

The single-ramp voltage generator of Fig. 6-10 generates an output voltage that depends on time, in accordance with Eq. (6-9). If V_o is known, the time that has elapsed since the capacitor began charging can be calculated. This principle is used in Fig. 6-11(a) to make an interval timer. Op amp A is a ramp generator that generates a negative-going ramp when the control switch is thrown to *start*. Op amp B is a comparator that monitors the ramp voltage with its negative input. An adjustable negative reference voltage is applied to the $(+)$ input of op amp B. When the ramp voltage crosses the reference voltage, the output voltage of the comparator snaps to $+V_{sat}$. This action is shown by the waveshapes in Fig. 6-11(b). R_L can be a sensitive relay or other power-control device that would, for example, activate an alarm or turn off a lamp. The diode in series with R_L conducts current only when V_o goes positive.

6-7.2 Circuit Analysis

The circuit is analyzed beginning with an example.

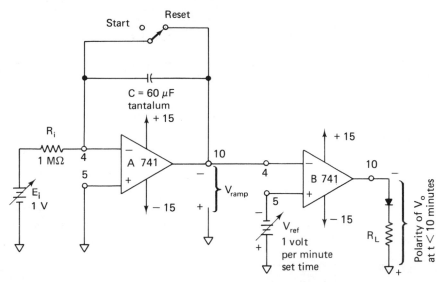

(a) Single-ramp generator A and comparator B make an interval timer

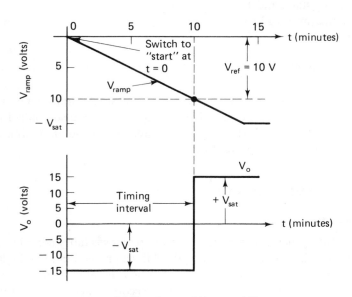

(b) Waveforms of V_{ramp} and V_0

Figure 6-11 Adjustable minute timer.

Example 6-8

(a) Find the rate at which V_o drops in volts per second for the single-ramp generator in Fig. 6-11(a). (b) Convert this *ramp-down* rate to volts per minute.
Solution. (a) Rewrite Eq. (6-9) to give volts per second:

$$\text{volts per second} = \frac{V_{\text{ramp}}}{t} = -\frac{E_i}{R_i C}$$

Calculate $R_i \times C = 1\ \text{M}\Omega \times 60\ \mu\text{F} = 60\ \text{s}$. Substituting into Eq. (6-9) for E_i and $R_i C$, we obtain

$$\text{volts per second} = \frac{V_{\text{ramp}}}{t} = -\frac{1\ \text{V}}{60\ \text{s}}$$

(b) Convert seconds to minutes:

$$\frac{V_{\text{ramp}}}{t} = -\frac{1\ \text{V}}{60\ \text{s}} \times \frac{60\ \text{s}}{1\ \text{min}} = -\frac{1\ \text{V}}{\text{min}}$$

We learn from Example 6-8 that E_i can control the rate at which V_{ramp} drops. For example, if E_i is doubled to 2 V, the ramp will drop twice as fast, at a rate of 2 V/min.

Example 6-9

If $V_{\text{ref}} = -10\ \text{V}$ in Fig. 6-11, how much time will elapse between the time the switch is opened (start) and V_o goes to $+V_{\text{sat}}$.
Solution. The comparator will switch when $V_{\text{ramp}} = V_{\text{ref}}$. The timing interval can be found from

$$\text{timing interval} = \frac{V_{\text{ref}}}{V_{\text{ramp}}/t} = \frac{10\ \text{V}}{1\ \text{V/min}} = 10\ \text{min}$$

Now that a comparator has been connected to a single-ramp generator, it is natural to ask whether the output of the comparator can also control the input of the ramp generator. The answer is yes, and such an op amp circuit is used to build a triangular-wave generator.

6-8 TRIANGULAR-WAVE GENERATOR

6-8.1 Introduction

A triangular-wave generator that requires two op amps is fairly complicated to analyze all at once. Understanding is simplified if we proceed in three logical steps. First, show how a basic triangular wave can be generated manually with one op amp, a resistor, a capacitor, and a switch. Second, select a comparator to replace the manual operation of the switch. Third, put the comparator and basic triangular-wave generator together. We begin with applying the principles of the single-ramp generator studied in Section 6-3.

6-8.2 Basic Operation

A manually controlled triangular-wave generator can be made by adding one switch and another dc control voltage to the single-ramp generator of Fig. 6-10. The resulting circuit is shown in Fig. 6-12(a). When the control switch is

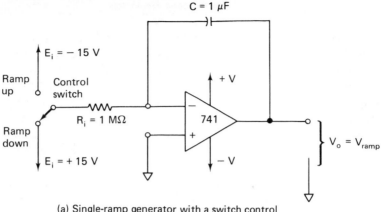

(a) Single-ramp generator with a switch control

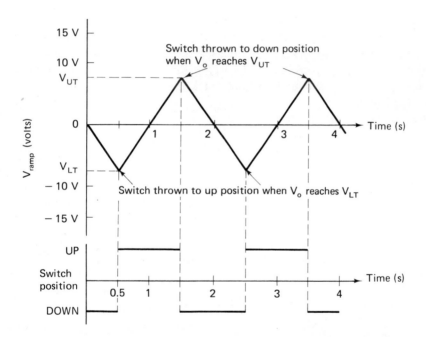

(b) Timing diagram showing how V_o varies with switch position

Figure 6-12 Manually controlled triangular-wave generator.

in the up position, E_i is -15 V and V_o ramps up. When the control switch is in the down position, E_i is $+15$ V and V_o ramps down. The rate of change of the ramp voltage is found from Eq. (6-9), for $E_i = +15$ V as

$$\frac{V_o}{t} = -\frac{E_i}{R_i C} = -\frac{15 \text{ V}}{1 \text{ M}\Omega \times 1 \text{ }\mu\text{F}} = -\frac{15 \text{ V}}{\text{s}}$$

For $E_i = -15$ V, $V_o/t = 15$ V/s.

To see how the ramp voltages are converted to a triangular wave, refer to Fig. 6-12(b). At time $t = 0$, power is turned on with the switch in the "ramp-down" position. E_i is positive, so V_o ramps down at -15 V/s. When V_o reaches a selected *lower-threshold voltage*, V_{LT}, the control switch is thrown to the "ramp-up" position. E_i is now -15 V and V_o ramps up at $+15$ V/s. When V_o reaches a selected upper-threshold voltage, V_{UT}, the control switch is thrown back to the "ramp-down" position. From then on the control switch position must be changed every time the ramp voltage V_o crosses one of the threshold voltages.

To make operation of the control switch automatic, we can replace it with a comparator. The type of comparator needed for the triangular wave is shown in Fig. 6-13, where its input is connected to the output of the ramp generator. (For a reference to this comparator, see Fig. 4-7.)

Operation of the triangle-wave generator of Fig. 6-13(a) is analyzed by reference to time B in Fig. 6-13(b). When V_{ramp} crosses V_{UT}, the comparator output snaps positive to $+V_{sat}$. This causes the ramp generator's output to ramp down until V_{ramp} drops just below V_{LT} at time C. Then V_{comp} snaps negative, causing the ramp generator to ramp back up to V_{UT}. This completes one cycle of the triangular wave.

The design-analysis equations are given by

$$V_{LT} = -\frac{+V_{sat}}{n} \tag{6-10a}$$

$$V_{UT} = -\frac{-V_{sat}}{n} \tag{6-10b}$$

$$V_H = \frac{+V_{sat} - (-V_{sat})}{n} \tag{6-10c}$$

If the magnitudes of $+V_{sat}$ and $-V_{sat}$ are equal, the frequency of oscillation is found from

$$f = \frac{n}{4R_i C} \tag{6-10d}$$

Example 6-10

For the triangular-wave generator of Fig. 6-13, find (a) V_{LT}; (b) V_{UT}; (c) frequency of oscillation. Assume that $\pm V_{sat} = \pm 15$ V.
Solution. (a) Since $R = 10$ kΩ and $nR = 20$ kΩ, $n = nR/R = 20$ kΩ$/10$ kΩ $= 2$. From Eq, (6-10a),

$$V_{LT} = -\frac{15}{2} = -7.5 \text{ V}$$

(b) From Eq. (6-10b),

$$V_{UT} = -\frac{-15}{2} = +7.5 \text{ V}; \qquad \text{therefore,} \quad V_H = 15 \text{ V}$$

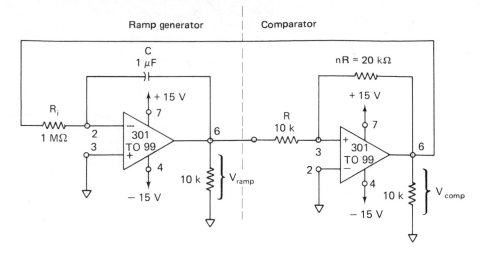

(a) Single-ramp generator plus comparator

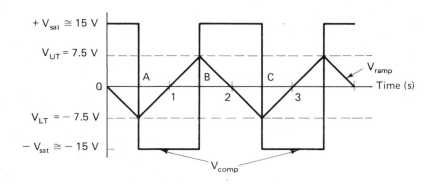

(b) Available output voltages

Figure 6-13 Triangle-wave generator.

(c) From Eq. (6-10d),

$$f = \frac{2}{4 \times 10^6 \times 1 \times 10^{-6}} = 0.5 \text{ Hz}$$

Note that if n is made equal to 4, then $f = R_i C$.

6-9 SAWTOOTH-WAVE GENERATOR

6-9.1 Introduction

The single-ramp generator in Fig. 6-11 was used to make a minute timer. With a slight modification, we can make a sawtooth-wave generator by continually resetting the timer. For example, in Fig. 6-14(a) suppose that the

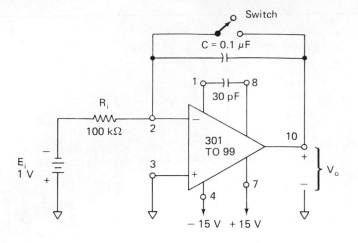

(a) Single-ramp generator with capacitor discharge switch

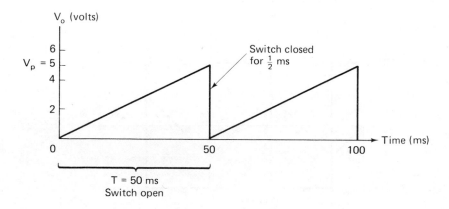

(b) Sawtooth wave shape of V_o, if the switch in (a) is closed
briefly every time V_o reaches 5 V

Figure 6-14 Manually controlled sawtooth-wave generator.

switch is open. V_o will rise at a rate of

$$\frac{V_o}{t} = -\frac{E_i}{R_i C} = -\frac{-1\ \text{V}}{(100\ \text{k}\Omega)(0.1\ \mu\text{F})} = \frac{1\ \text{V}}{10\ \text{ms}}$$

This expression shows that in every 10-ms time interval, V_o will go more positive
by 1 V. If the switch is closed, the capacitor will quickly discharge through the
short circuit, and V_o will drop to 0 V. If the switch is opened, the capacitor
begins again to charge and V_o rises. By quickly closing and opening the switch
every time V_o rises to a peak voltage V_p (for example, 5 V), the sawtooth waveshape shown in Fig. 6-14(b) would be generated. The frequency f of the sawtooth-wave generator is found from

$$f = \frac{E_i}{R_i C} \times \frac{1}{V_p} \qquad \text{(6-11a)}$$

The period of oscillation is

$$T = \frac{1}{f} \qquad \text{(6-11b)}$$

Example 6-11

In Fig. 6-14, what is (a) the frequency; (b) the period of oscillation?
Solution. (a) By Eq. (6-11a),

$$f = \frac{1 \text{ V}}{10 \text{ ms}} \times \frac{1}{5 \text{ V}} = 20 \text{ Hz}$$

(b) By Eq. (6-11b),

$$T = \frac{1}{20 \text{ Hz}} = 50 \text{ ms}$$

To generate the sawtooth wave automatically, we need a device or circuit that will do four jobs in the following order:

1. Sense when the capacitor voltage reaches a desired peak value V_p.
2. Place a short circuit across the capacitor.
3. Finally, sense when the capacitor is almost completely discharged.
4. Remove the short circuit.

There is such a device, and it is inexpensive. It is the *programmable unijunction transistor* (PUT).

6-9.2 Programmable Unijunction Transistor

The PUT is a three-terminal device that acts as a voltage-sensitive switch. Its schematic diagram is shown in Fig. 6-15(a). The switch terminals are labeled anode A and cathode K. Current flows only from anode to cathode (as indicated by the arrow).

Normally, the switch terminals act as an open circuit. To make the switch terminals act as a short circuit whenever the desired voltage level V_p is reached, the PUT must compare this changing voltage level against a fixed reference voltage. Therefore, a third terminal is needed where the reference voltage is applied. This third terminal is called the *gate*, G. Apply the voltage you want to sense to the anode terminal. When the anode voltage becomes more positive than the gate voltage (by a few tenths of a volt), the PUT's anode and cathode suddenly act as a short circuit. They remain a short circuit independently of the gate terminal until the anode-to-cathode current drops below the PUT's *holding current*, I_H. (I_H is typically a few milliamperes.) Then the anode and cathode terminals abruptly act as an open circuit. The cathode K of a PUT

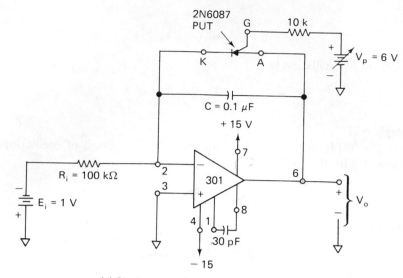

(a) Single-ramp generator with programmable
unijunction transistor (PUT)

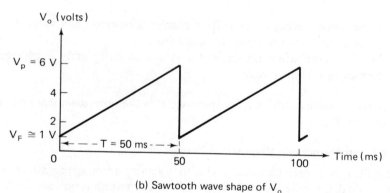

(b) Sawtooth wave shape of V_o

Figure 6-15 Sawtooth-wave generator.

should be connected to ground, as it is by the virtual ground at pin 2 in Fig.
6-15(a). How the PUT's characteristics are used in the sawtooth-wave generator
is shown in Section 6-9.3.

6-9.3 Sawtooth-Wave-Generator Operation

The simplest sawtooth-wave generator is shown in Fig. 6-15(a). The desired
peak voltage of the sawtooth wave, V_p, is applied to the gate terminal of the
PUT. E_i drives V_o positive at 1 V/10 ms until V_o exceeds V_p by a few tenths of
a volt. The PUT's anode and cathode terminals then abruptly act as a short
circuit and discharge capacitor C to about 1 V. When capacitor discharge
current becomes less than the PUT's holding current, the PUT's anode and

cathode terminals abruptly act as an open circuit. Unfortunately, the PUT does not act as an ideal short circuit because its A–K terminal voltage drops only to a *forward voltage* V_F equal to about 1 V, not 0 V. But by increasing V_p by about 1 V to compensate for V_F, we can generate a sawtooth wave with a frequency of 20 Hz as in Fig. 6-15(b). Note that V_o now varies between $V_F \approx$ 1 V and $V_p \approx 6$ V. The frequency of oscillation is expressed by

$$f = \frac{E_i}{R_i C} \times \frac{1}{V_p - 1} \tag{6-12}$$

Equation (6-12) will be studied in more detail in Section 6-10.

6-10   VOLTAGE-TO-FREQUENCY CONVERTERS

6-10.1   Voltage-Controlled Oscillator

Equation (6-12) shows that the frequency of a sawtooth-wave generator depends on two factors:

1. How fast V_o rises in volts per second, $E_i/R_i C$
2. The value to which V_o can rise, $(V_p - 1)$ V

Therefore, frequency can be controlled by either voltage E_i or V_p. This is shown in the next two examples.

Example 6-12

If E_i is doubled to 2 V in Fig. 6-15(a), what is the new frequency of oscillation?
Solution. From Eq. (6-12),

$$f = \frac{2 \text{ V}}{(100 \text{ k}\Omega)(0.1 \ \mu\text{F})} \times \frac{1}{(6 - 1) \text{ V}} = 40 \text{ Hz}$$

The frequency is doubled because V_o rises twice as fast.

Example 6-13

If V_p is approximately halved to 3.5 V in Fig. 6-15(a), what is the new frequency of oscillation? $E_i = 1$ V.
Solution. From Eq. (6-12),

$$f = \frac{1 \text{ V}}{(100 \text{ k}\Omega)(0.1 \ \mu\text{F})} \times \frac{1}{(3.5 - 1) \text{ V}} = \frac{1 \text{ V}}{0.010 \text{ s}} \times \frac{1}{2.5 \text{ V}} = 40 \text{ Hz}$$

The frequency is doubled because V_o now has to rise only to 2.5 V rather than to 5 V.

Both examples show that our sawtooth-wave generator can convert a voltage E_i or V_p to a frequency; therefore, it is a voltage-to-frequency converter.

6-10.2 Frequency Modulation and Frequency Shift Keying

Examples 6-12 and 6-13 indicate one way of achieving *frequency modulation* (FM). Thus, if the amplitude of E_i varies, the frequency of the sawtooth oscillator will be changed or modulated. If E_i is keyed between two voltage levels, the sawtooth oscillator changes frequencies. This type of application is called *frequency shift keying* (FSK) and is used for data transmission. These two preset frequencies correspond to "0" and "1" states (commonly called *space* and *mark*) in binary.

6-11 SINE-WAVE OSCILLATOR

6-11.1 Oscillator Theory

A basic sine-wave oscillator should generate a sine wave at only one frequency. To do this, assume that different frequencies are present at some starting point in a circuit. The frequencies are passed through a frequency-selection network where they are reduced in amplitude and exit with different phase angles. Next they are passed through an amplifier to replace the amplitude lost in the selection circuit. We hope that only *one* of the frequencies exiting from the amplifier will be an exact copy of one of the original frequencies both in magnitude and phase angle. Then, by connecting the amplifier output back to the starting point, we have an oscillator that will oscillate at only this one frequency.

To explore this principle in detail, refer to Fig. 6-16(a), in which three frequencies are considered. E_o is the desired oscillating frequency. E_H is a higher frequency, and E_L is a lower frequency. At the output of the frequency selector, E'_o has the same phase angle as input E_o, although its peak value is reduced by one-third. (Note the different voltage scales between input and output.) Output E'_H is not in phase with E_H, nor is E'_L in phase with E_L.

Visualize all three outputs of the frequency selector applied to a noninverting amplifier with a gain of 3. As shown in Fig. 6-16(b), only one frequency, E_o, will emerge from the amplifier with exactly the same amplitude and, more important, exactly the same phase as the original E_o.

By connecting the amplifier output back to the input of the frequency selector, we close a loop. Now a signal of only one frequency can be transmitted completely around the loop without changing its magnitude or phase angle. To put it another way, only one frequency has a loop gain of 1. All other signals will be attenuated after each pass around the loop and experience a phase shift that will act to dampen them out.

One final point: Where did E_o come from in the first place? The answer is

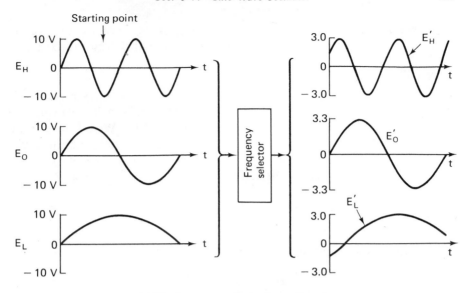

(a) The frequency selector transmits only one
frequency, E_o, with no change in phase angle

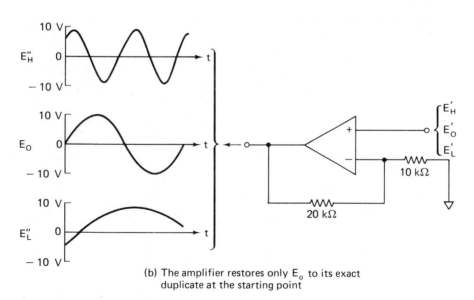

(b) The amplifier restores only E_o to its exact
duplicate at the starting point

Figure 6-16   Sine-wave oscillator theory.

that either a transient voltage generated by turning the power on or the inevitable presence of noise introduces a number of frequencies within the loop.
Among them will be the one oscillating frequency E_o that will see a loop gain
of 1. Actually, the loop gain should be slightly greater than 1 so that E_o can

build up. As will be shown in Section 6-11.3, we must make provision to prevent E_o from building up indefinitely and forcing the amplifier into saturation.

6-11.2 Setting Up an Oscillator

Figure 6-17 presents a practical circuit to illustrate the ideas presented in Section 6-11.1. E_i is a voltage from an audio oscillator. The RC network is a Wein bridge type of frequency selector. The output of the Wein bridge, E', is applied to the noninverting amplifier. The gain of the amplifier is adjusted with the 50-kΩ pot. If the frequency of E_i is varied, only one frequency will be developed at E' having a phase shift of $0°$. This frequency, f_0, is calculated from

$$f_0 = \frac{1}{2\pi RC} \tag{6-13}$$

Attenuation of the frequency selector at f_0 is one-third, or

$$E' = \tfrac{1}{3}E_i \qquad \text{at} f_0$$

By adjusting the amplifier's gain (50-kΩ pot), we can amplify E' by a gain of 3 so that the amplifier output V_o is precisely equal to E_i for amplifier gain $= 3$, and frequency $= f_0$.

Next, monitor V_o with a CRO and flip the switch from "test" to "oscillate." This action closes the loop, which has a net gain of 1 ($\tfrac{1}{3} \times 3 = 1$). One of two events will then be seen on the CRO at V_o. Either V_o will switch from $+V_{sat}$ to $-V_{sat}$ at the frequency f_0, or V_o will not show any oscillation and lock at $+V_{sat}$ or $-V_{sat}$. By adjusting the 50-kΩ potentiometer, you will see briefly a

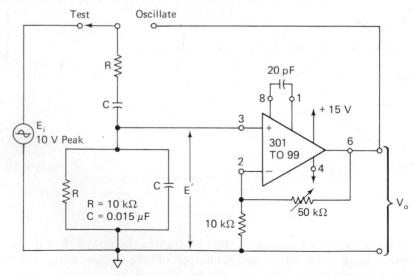

Figure 6-17 Test circuit to find the frequency of oscillation $f_o = 1/(2\pi RC)$.

nice sine wave, whose amplitude will either decay or build up. The oscillator is almost working but needs some amplitude control. One technique for introducing amplitude control is given in Section 6-11.3.

6-11.3 Wein Bridge Oscillator

Experience with the test circuit of Fig. 6-17 showed that the output voltage V_o could increase without limit once oscillation began. What is needed is a circuit that will sense the amplitude of the output voltage and reduce amplifier gain when the voltage exceeds a specified level. We add a back-to-back zener diode and one resistor to do this job in the practical Wein bridge oscillator of Fig. 6-18. When the oscillator's output voltage increases above the zener voltage, one zener or the other (depending on the polarity of V_o) breaks down. The zener then shunts the 10-kΩ resistor to reduce the gain of the amplifier and prevent V_o from being driven to $\pm V_{sat}$. The 25-kΩ resistor allows adjustment of V_o from peak values of about $1.5V_Z \approx 8$ V to $\pm V_{sat}$. The resulting sine-wave output has very little distortion. For best results, the oscillator's output should be connected to a voltage follower to avoid undue loading.

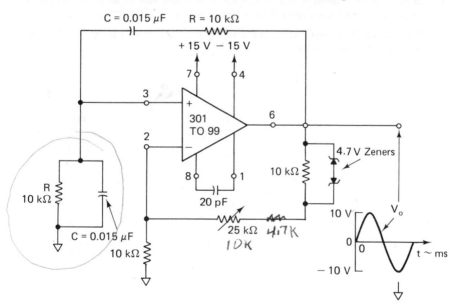

Figure 6-18 Practical Wein bridge sine-wave oscillator, $f_o \cong 1$ kHz.

PROBLEMS

6-1. In Example 6-1, saturation voltages are ±10 V. Find V_{UT} and V_{LT}.

6-2. Find the period of the multivibrator in Fig. 6-1 if $R_f = 10$ kΩ and $C = 0.01$ μF.

6-3. Find the frequency of oscillation in Problem 6-2 above.

6-4. E_i must go negative by what value in Fig. 6-4(a) to trigger the one-shot into its timing state?

6-5. What is another name for a one-shot?

6-6. In Fig. 6-4, $R_f = 10$ kΩ and $C = 0.1$ μF. Calculate the duration of the output pulse.

6-7. If D_1 and D_2 are reversed in Fig. 6-4, what is the effect on circuit operation?

6-8. Why would you add diode D_D and resistor R_D to the circuit of Fig. 6-5?

6-9. If V_{UT} and V_{LT} are adjusted to $+4$ V and -4 V respectively in Example 6-5, what is the new frequency of oscillation?

6-10. If I is doubled in Example 6-6, will the capacitor voltage be doubled or halved in 5 seconds?

6-11. If E_i is doubled in Fig. 6-10, will the ramp-down rate double or halve?

6-12. Find the time interval in Fig. 6-11 if $V_{ref} = -5$ V.

6-13. Find the oscillating frequency in Fig. 6-13 if (a) only C is doubled to 2 μF; (b) only nR is doubled to 40 kΩ; (c) if only R_i is doubled to 2 kΩ.

6-14. In the sawtooth generator of Fig. 6-15, how could you generate a frequency of 10 Hz by (a) changing only R_i; (b) changing only C; (c) changing only V_p?

6-15. If C is doubled to 0.03 μF in the sine-wave oscillator of Fig. 6-18, what is the new frequency of oscillation?

op amps with diodes

ᴨᵧᴨ

7

7-0 INTRODUCTION TO PRECISION RECTIFIERS

The major limitation of ordinary silicon diodes is that they cannot rectify voltages below 0.6 V. For example, Fig. 7-1(a) shows that V_o does not respond to positive inputs below 0.6 V in a half-wave rectifier built with an ordinary silicon diode. Figure 7-1(b) shows the waveforms for a half-wave rectifier built with an ideal diode. An output voltage occurs for all positive input voltages, even those below 0.6 V. A circuit that acts like an ideal diode can be designed using an op amp and two ordinary diodes. The result is a powerful circuit capable of rectifying input signals of only a few millivolts.

The low cost of this equivalent ideal diode circuit allows it to be used routinely for many applications. They can be grouped loosely into the following classifications: linear half-wave rectifiers and precision full-wave rectifiers.

1. *Linear half-wave rectifiers.* The linear half-wave rectifier circuit delivers an output that depends on the magnitude *and polarity* of the input voltage. The output is inverted with respect to the input.

   It will be shown that the linear half-wave rectifier circuit can be modified to perform a variety of signal-proccessing applications. The linear half-wave rectifier is also called a *precision half-wave rectifier* and acts as an ideal diode.

2. *Precision full-wave rectifiers.* The precision full-wave rectifier circuit delivers an output proportional to the magnitude but *not* the polarity

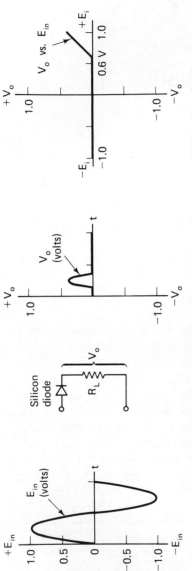

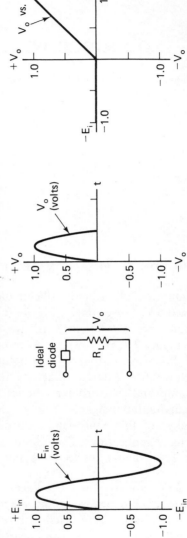

(a) Real diodes cannot rectify small ac voltages because of the diode's 0.6-V voltage drop

(b) A linear or precision half-wave rectifier circuit precisely rectifies any ac signal regardless of amplitude and acts as an ideal diode

Figure 7-1 The ordinary silicon diode requires about 0.6 V of forward bias in order to conduct. Therefore, it cannot rectify small ac voltages. A precision half-wave rectifier circuit overcomes this limitation.

of the input. For example, the output can be positive at 2 V for inputs of either $+2$ V or -2 V. Since the absolute value of $+2$ V and -2 V is equal to $+2$ V, the precision full-wave rectifier is also called an *absolute-value circuit*.

Applications for both linear half-wave and precision full-wave rectifiers include:

1. Detection of amplitude-modulated signals
2. Dead-zone circuits
3. Precision bound circuits or *clippers*
4. Current switches
5. Waveshapers
6. Peak-value indicators
7. Sample-and-hold circuits
8. Absolute-value curcuits
9. Averaging circuits
10. Signal polarity detectors
11. Ac-to-dc converters

7-1 LINEAR HALF-WAVE RECTIFIERS

7-1.1 Introduction

Linear half-wave rectifier circuits transmit only one-half cycle of a signal and eliminate the other by *bounding* the output to zero volts. The input half-cycle that is transmitted can be either inverted or noninverted. It can also experience gain, attenuation, or remain unchanged in magnitude, depending on the choice of resistors and placement of diodes in the op amp circuit.

7-1.2 Inverting, Linear Half-Wave Rectifier, Positive Output

The inverting amplifier is converted into an ideal (linear precision) half-wave rectifier by adding two diodes as shown in Fig. 7-2. When E_i is positive in Fig. 7-2(a), diode D_1 conducts causing the op amp's output voltage, V_{OA}, to go negative by one diode drop ($\cong 0.6$ V). This forces diode D_2 to be reverse biased. The circuit's output voltage V_o equals zero because input current I flows through D_1. For all practical purposes, no current flows through R_F and therefore $V_o = 0$.

Note the load is modeled by a resistor R_L and must always be resistive. If the load is a capacitor, inductor, voltage, or current source, then V_o will *not* equal zero.

In Fig. 7-2(b), negative input E_i forces the op amp output V_{OA} to go positive. This causes D_2 to conduct. The circuit then acts like an inverter since $R_F = R_i$

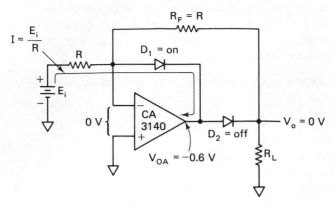

(a) Output V_o is bound at 0 V for all
    positive input voltages

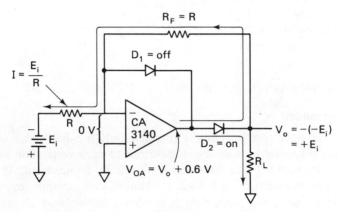

(b) Output V_o is positive and equal to the
    magnitude of E_i for all negative inputs

Figure 7-2 Two diodes convert an inverting amplifier into a positive-output, inverting, linear (ideal) half-wave rectifier. Output V_o is positive and equal to the magnitude of E_i for negative inputs and V_o equals 0 V for all positive inputs. Diodes are IN914 or IN4154.

and $V_o = -(-E_i) = +E_i$. Since the $(-)$ input is at ground potential, diode D_1 is reverse biased. Input current is set by E_i/R_i and gain by $-R_F/R_i$. Remember that this gain equation applies only for negative inputs, and V_o can only be positive or zero.

Circuit operation is summarized by the waveshapes in Fig. 7-3. V_o can only go positive in a linear response to negative inputs. The most important property of this linear half-wave rectifier will now be examined. An ordinary silicon diode or even a hot-carrier diode requires a few tenths of volts to become

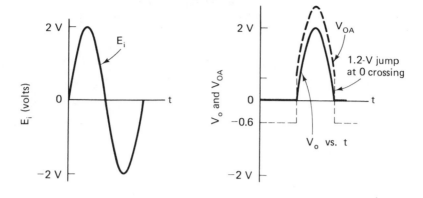

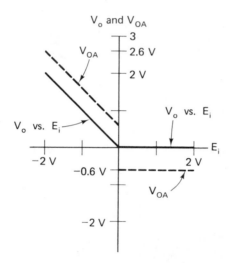

Figure 7-3 Input, output, and transfer characteristics of a positive-output, ideal, inverting half-wave rectifier.

forward biased. Any signal voltage below this threshold voltage cannot be rectified. However, by connecting the diode in the feedback loop of an op amp, the threshold voltage of the diode is essentially eliminated. For example, in Figure 7-2(b) let E_i be a low voltage of -0.1 V. E_i and R_i convert this low voltage to a current that is conducted through D_2. V_{OA} goes to whatever voltage is required to supply the necessary diode drop plus the voltage drop across R_F. Thus millivolts of input voltage can be rectified since the diode's forward bias is supplied automatically by the negative feedback action of the op amp.

Finally, observe the waveshape of op amp output V_{OA} in Fig. 7-3. When E_i crosses 0 V (going negative), V_{OA} jumps quickly from -0.6 V to $+0.6$ V as it switches from supplying the drop for D_2 to supplying the drop for D_1. This jump can be monitored by a differentiator to indicate the zero crossing. During the jump time the op amp operates open loop.

7-1.3   Inverting Linear Half-Wave Rectifier, Negative Output

The diodes in Fig. 7-2 can be reversed as shown in Fig. 7-4. Now only positive input signals are transmitted and inverted. The output voltage V_o equals 0 V for all negative inputs. Circuit operation is summarized by the plot of V_o and V_{OA} versus E_i in Fig. 7-4(b).

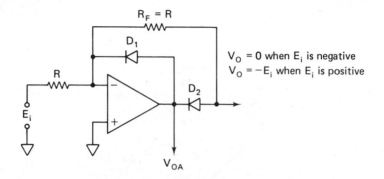

$V_O = 0$ when E_i is negative
$V_O = -E_i$ when E_i is positive

(a) Inverting linear half-wave rectifier: negative output

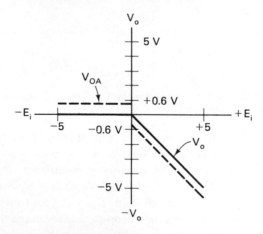

(b) Transfer characteristic V_o vs. E_i

Figure 7-4   Reversing the diodes in Fig. 7-2 gives an inverting linear half-wave rectifier. This circuit transmits and inverts only positive input signals.

7-1.4 Signal Polarity Separator

The circuit of Fig. 7-5 is an expansion of the circuits in Figs. 7-2 and 7-4. When E_i is positive in Fig. 7-5(a), diode D_1 conducts and an output is obtained only on output V_{o1}. V_{o2} is bound at 0 V. When E_i is negative, D_2 conducts, $V_{o2} = -(-E_i) = +E_i$, and V_{o1} is bound at 0 V. This circuit's operation is summarized by the waveshapes in Fig. 7-6.

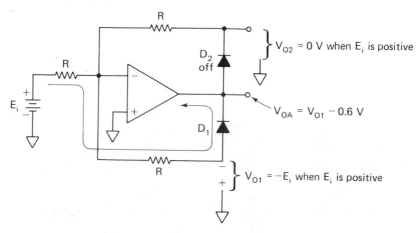

(a) When E_i is positive, V_{O1} is negative and V_{O2} is bound at 0 V

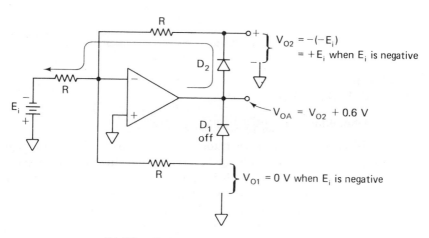

(b) When E_i is negative, $V_{O1} = 0$ V and V_{O2} goes positive

Figure 7-5 This circuit inverts and separates the polarities of input signal E_i. A positive output at V_{O2} indicates that E_i is negative and a negative output at V_{O1} indicates that E_i is positive. These outputs should be buffered.

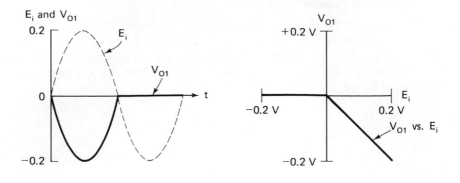

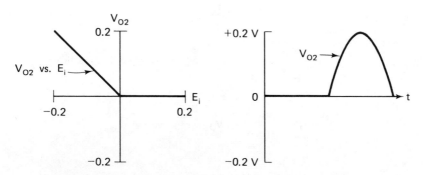

Figure 7-6   Input and output voltages for the polarity separator of Fig. 7-5.

7-2   PRECISION RECTIFIERS—
THE ABSOLUTE-VALUE CIRCUIT

7-2.1   Introduction

The precision full-wave rectifier transmits one polarity of the input signal and inverts the other. Thus both half-cycles of an alternating voltage are transmitted but are converted to a single polarity at the circuit's output. The precision full-wave rectifier can rectify input voltages with millivolt amplitudes.

This type of circuit is useful to prepare signals for multiplication, averaging, or demodulation. The characteristics of an ideal precision rectifier are shown in Fig. 7-7.

The precision rectifier is also called an *absolute-value* circuit. The absolute value of a number (or voltage) is equal to its magnitude regardless of sign. For example, the absolute values of $|+2|$ and $|-2|$ are equal at 2. The symbol $|*|$ means "absolute value of." Figure 7-7 shows that the output equals the absolute value of the input. In a precision rectifier circuit the output is either negative or positive, depending on how the diodes are installed.

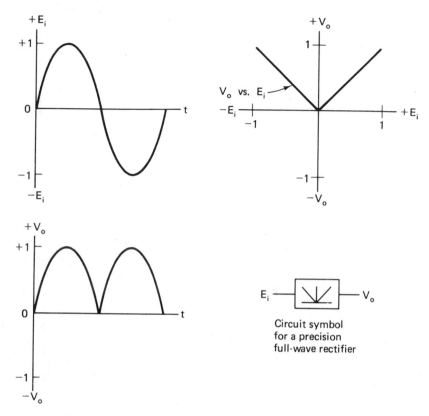

Figure 7-7 The precision full-wave rectifier fully rectifies input voltages, including those with values less than a diode threshold voltage.

7-2.2 Types of Precision Full-Wave Rectifiers

Three types of precision rectifiers will be presented. The first is inexpensive because it uses two op-amps, two diodes, and five *equal* resistors. Unfortunately, it does not have high input resistance, so a second type is given that does have high input resistance but requires resistors that are precisely proportioned but *not* all equal. Neither type has a summing node at virtual ground potential, so a third type will be presented in Section 7-4.2 to allow averaging.

7-2.3 Full-Wave Precision Rectifier with Equal Resistors

The first type of precision full-wave rectifier or absolute-value circuit is shown in Fig. 7-8. This circuit uses equal resistors and has an input resistance equal to R. Figure 7-8(a) shows current directions and voltage polarities for positive input signals. Diode D_P conducts so that both op amps A and B act as inverters, and $V_o = +E_i$.

Figure 7-8(b) shows that for negative input voltages, diode D_N conducts.

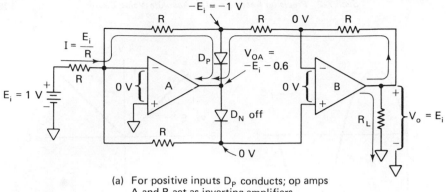

(a) For positive inputs D_P conducts; op amps A and B act as inverting amplifiers

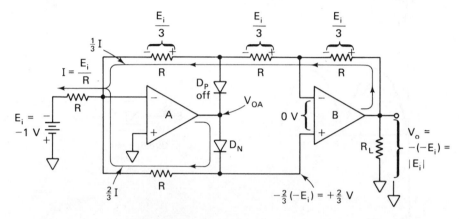

(b) For negative inputs, D_N conducts

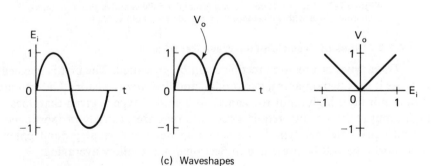

(c) Waveshapes

Figure 7-8 Absolute-value circuit or precision full-wave rectifier. $V_o = |E_i|$.

Input current I divides as shown, so that op amp B acts as an inverter. Thus output voltage V_o is positive for either polarity of input E_i and V_o is equal to the absolute value of E_i.

Waveshapes in Fig. 7-8(c) show that V_o is always of positive polarity and equal to the absolute value of the input voltage. To obtain negative outputs for either polarity of E_i simply reverse the diodes.

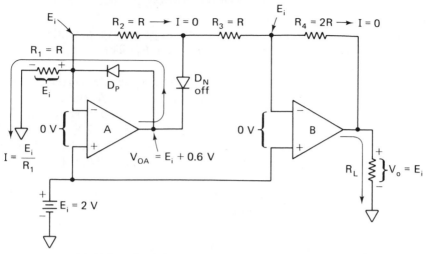

(a) Voltage levels for positive inputs: $V_o = +E_i$ for all positive E_i

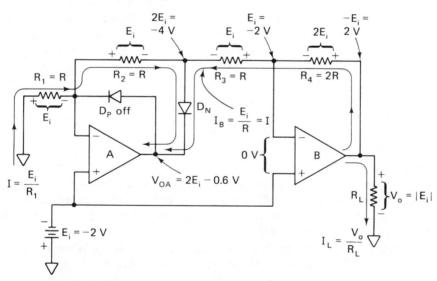

(b) Voltage levels for negative inputs: $V_o = -(-E_i) = |E_i|$

Figure 7-9 Precision full-wave rectifier with high input impedance. $R = 10$ kΩ, $2R = 20$ kΩ.

7-2.4 High-Impedance Precision Full-Wave Rectifier

The second type of precision rectifier is shown in Fig. 7-9. The input signal is connected to the noninverting op amp inputs to obtain high input impedance. Figure 7-9(a) shows what happens for positive inputs. E_i and R_1 set the current through diode D_P. The $(-)$ inputs of both op amps are at a potential equal to E_i so that no current flows through R_2, R_3, and R_4. Therefore, $V_o = E_i$ for all positive input voltages.

When E_i goes negative in Fig. 7-9(b), E_i and R_1 set the current through both R_1 and R_2 to turn on diode D_N. Since $R_1 = R_2 = R$, the anode of D_N goes to $2E_i$ or $2(-2) = -4$ V. The $(-)$ input of op amp B is at $-E_i$. The voltage drop across R_3 is $2E_i - E_i$ or $(-4$ V$) - (-2) = -2$ V. This voltage drop and R_3 establishes a current I_3 through both R_3 and R_4 equal to the input current I. Consequently, $V_o = -E_i$ when E_i is negative. Thus V_o is always positive despite the polarity of E_i, so $V_o = |E_i|$.

Waveshapes for this circuit are the same as in Fig. 7-8(c). Note that the maximum value of E_i is limited by the negative saturation voltage of the op amps.

7-3 PEAK DETECTORS

In addition to precisely rectifying a signal, diodes and op amps can be interconnected to build a peak detector circuit. This circuit follows the voltage peaks of a signal and stores the highest value (almost indefinitely) on a capacitor. If a higher peak signal value comes along, this new value is stored. The highest peak voltage is stored until the capacitor is discharged by a mechanical or electronic switch. This peak detector circuit is also called a *follow-and-hold* or *peak follower*. We shall also see that reversing two diodes changes this circuit from a peak to a *valley follower*.

7-3.1 Positive Peak Follower and Hold

The peak *follower-and-hold* circuit is shown in Fig. 7-10. It consists of two op amps, two diodes, a resistor, a hold capacitor, and a reset switch. Op amp A is a precision half-wave rectifier that charges C only when input voltage E_i exceeds capacitor voltage V_C. Op amp B is a voltage follower whose output signal is equal to V_C. The follower's high input impedance does not allow the capacitor to discharge appreciably.

To analyze circuit operation, let us begin with Fig. 7-10(a). When E_i exceeds V_C, diode D_P is forward biased to charge hold capacitor C. As long as E_i is greater than V_C, C charges toward E_i. Consequently, V_C follows E_i as long as E_i exceeds V_C. When E_i drops below V_C, diode D_N turns on as shown in Fig. 7-10(b). Diode D_P turns off and disconnects C from the output of op amp A. Diode D_P must be a low-leakage-type diode or the capacitor voltage will discharge (droop). To minimize droop, op amp B should require small input bias currents and for that reason should be a metal-oxide-semiconductor (MOS) or bipolar-field-effect (BIFET) op amp.

Figure 7-11 shows an example of voltage waveshapes for a positive voltage follower-and-hold circuit. To reset the hold capacitor voltage to zero, connect a discharge path across it with a 2-kΩ resistor.

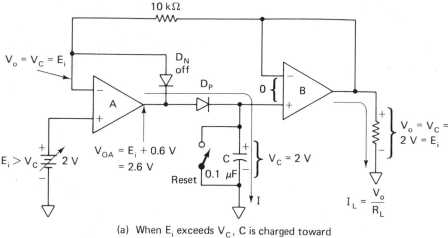

(a) When E_i exceeds V_C, C is charged toward
   E_i via D_P

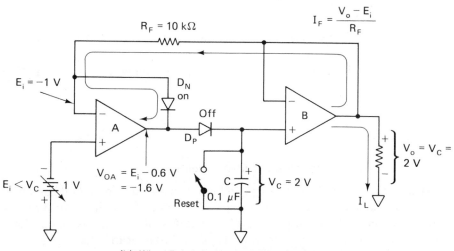

(b) When E_i is less than V_C, C holds its
   voltage at the highest previous value of E_i

Figure 7-10 Positive peak follower and hold or peak detector circuit. Op
amps are TL081 BIFETs.

7-3.2 Negative Peak Follower and Hold

When it is desired to hold the lowest or most negative voltage of a signal,
reverse both diodes in Fig. 7-10. For bipolar or negative input signals, V_o will
store the most negative voltage. It may be desired to monitor a positive voltage
and catch any negative dips of short duration. Simply jumper V_C to the positive
voltage to be monitored to load C with an equal positive voltage. Then when

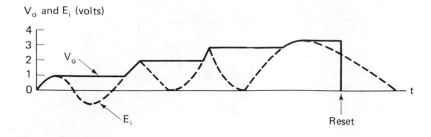

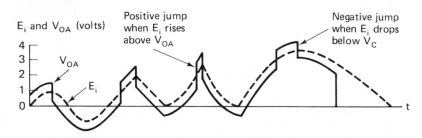

Figure 7-11 Waveshapes for the positive peak detector of Fig. 7-10(a).

the monitored voltage drops and recovers, V_C will follow the drop and store the lowest value.

7-4 AC-TO-DC CONVERTER

7-4.1 AC-to-DC Conversion or MAV Circuit

This section shows how to design and build an op amp circuit that computes the average value of a rectified ac voltage. This type of circuit is called an ac-to-dc converter. Since a full-wave rectifier circuit is also known as an absolute-value circuit and since an average value is also called a mean value, the ac-to-dc converter is also referred to as a *mean-absolute-value* (MAV) circuit.

To see how the MAV circuit is useful, refer to Fig. 7-12. A sine, triangle, and square wave are shown with equal maximum (peak) values. Therefore, a peak detector could not distinguish between them. The positive half-cycle and negative half-cycles are equal for each particular wave. Therefore, the average value of each signal is zero, so you could not distinguish one from another with an averaging circuit or device such as a dc volt meter. However, the MAV of each voltage is different (see Fig. 7-12).

The MAV of a voltage wave is approximately equal to its rms value. Thus an inexpensive MAV circuit can be used as a substitute for a more expensive rms calculating circuit.

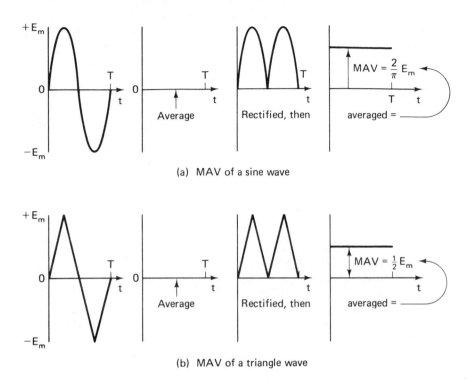

(a) MAV of a sine wave

(b) MAV of a triangle wave

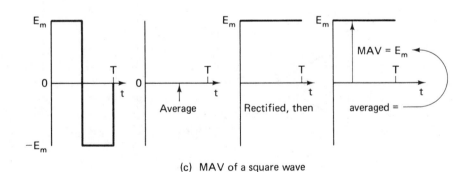

(c) MAV of a square wave

Figure 7-12   Mean absolute value of alternating sine, triangular, and square waves.

7-4.2 Precision Rectifier with Grounded Summing Inputs

To construct an ac-to-dc converter, we begin with the precision rectifier or absolute-value amplifier of Fig. 7-13. For positive inputs in Fig. 7-13(a), op amp A inverts E_i. Op amp B sums the output of A and E_i to give a circuit output

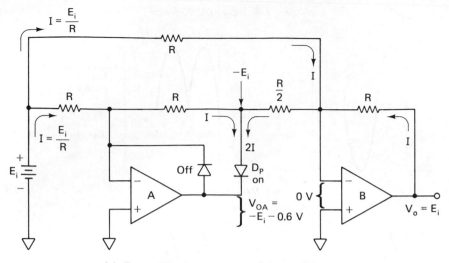

(a) For positive inputs, op amp A inverts E_i;
op amp B is an inverting adder, so that $V_o = E_i$

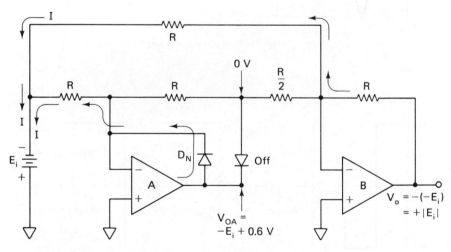

(b) For negative inputs, the output of A is
rectified to 0; op amp B inverts E_i, so
that $V_o = +E_i$

Figure 7-13 This absolute-value amplifier has both summing nodes at
ground potential during either polarity of input voltage. $R = 20 \text{ k}\Omega$.

of $V_o = E_i$. For negative inputs as shown in Fig. 7-13(b), op amp B inverts
$-E_i$ and the circuit output V_o is $+E_i$. Thus the circuit output V_o is positive and
equal to the rectified or absolute value of the input.

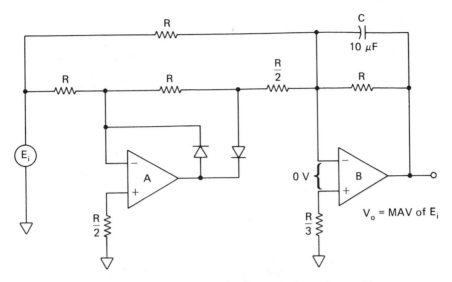

Figure 7-14 Ac-to-dc converter or mean-absolute-value amplifier.

7-4.3 AC-to-DC Converter

A large-value low-leakage capacitor (10-μF tantalum) is added to the absolute-value circuit of Fig. 7-13. The resultant circuit is the MAV amplifier or ac-to-dc converter shown in Fig. 7-14. Capacitor C does the averaging of the rectified output of op amp B. It takes about 50 cycles of input voltage before the capacitor voltage settles down to its final reading. If the waveshapes of Fig. 7-12 are applied to the ac-to-dc converter, its output will be the MAV of each wave.

7-5 DEAD-ZONE CIRCUITS

7-5.1 Introduction

Comparator circuits tell *if* a signal is below or above a particular reference voltage. In contrast with the comparator, a dead-zone circuit tells *by how much* a signal is below or above a reference voltage.

7-5.2 Dead-Zone Circuit with Negative Output

Analysis of a dead-zone circuit begins with the circuit of Fig. 7-15. A convenient regulated supply voltage $+V$ and resistor mR establish a reference voltage V_{ref}. V_{ref} is found from the equation $V_{ref} = +V/m$. As will be shown, the *negative* of V_{ref}, $-V_{ref}$, will establish the dead zone. In Fig. 7-15(a), current I is determined by $+V$ and resistor mR at $I = +V/mR$.

Diode D_N will conduct for all positive values of E_i, clamping V_{OA} and V_{OB}

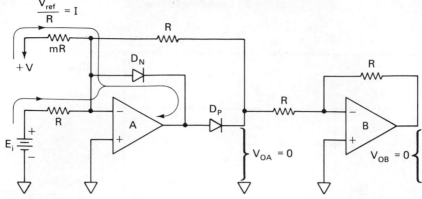

(a) $V_{ref} = +V/m$; V_{OA} and V_{OB} equal 0 for all positive values of E_i and all negative values of E_i above $-V_{ref}$

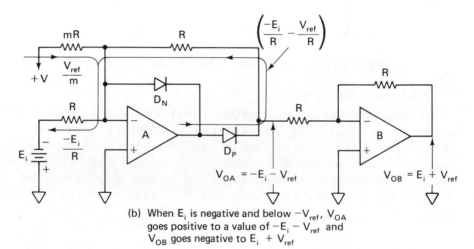

(b) When E_i is negative and below $-V_{ref}$, V_{OA} goes positive to a value of $-E_i - V_{ref}$ and V_{OB} goes negative to $E_i + V_{ref}$

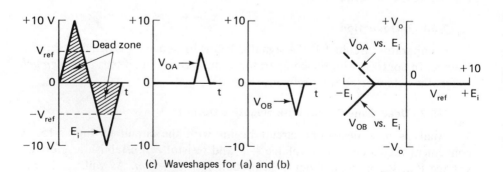

(c) Waveshapes for (a) and (b)

Figure 7-15 The dead-zone circuit output V_{OB} eliminates all portions of the signal above $-V_{ref} \cdot V_{ref} = +V/m$.

to 0 V. Therefore, all positive inputs are eliminated from affecting the output. In order to get any output at V_{OA}, E_i must go negative, as shown in Fig. 7-15(b). Diode D_P will conduct when the loop current E_i/R through E_i exceeds the loop current V/mR through resistor mR.

The value of E_i necessary to turn on D_P in Fig. 7-15(b) is equal to $-V_{ref}$. This conclusion is found by equating the currents

$$-\frac{E_i}{R} = \frac{+V}{mR}$$

and solving for E_i:

$$E_i = -\frac{+V}{m} = -V_{ref} \tag{7-1a}$$

where

$$V_{ref} = \frac{+V}{m} \tag{7-1b}$$

Thus all values of E_i above $-V_{ref}$ will lie in a dead zone where they will not be transmitted. When E_i is below $-V_{ref}$, E_i and V_{ref} will be added and inverted at V_{OA}, then reinverted at V_{OB}. Circuit operation is summarized by the waveshapes in Fig. 7-15(c) and the following example.

Example 7-1

In the circuit of Fig. 7-15, $+V = +15$ V, $mR = 30$ kΩ, and $R = 10$ kΩ, so that $m = 3$. Find (a) V_{ref}; (b) V_{OA} when $E_i = -10$ V; (c) V_{OB} when $E_i = -10$ V.

Solution. (a) From Eq. (7-1b), $V_{ref} = +15$ V$/3 = 5$ V.

(b) V_{OA} and V_{OB} will equal zero for all values of E_i above $-V_{ref} = -5$ V, from Eq. (7-1a). Therefore, $V_{OA} = -E_i - V_{ref} = -(-10) - 5$ V $= +5$ V.

(c) Op amp B inverts the output of V_{OA} so that $V_{OB} = -5$ V. Thus V_{OB} indicates how much E_i goes below $-V_{ref}$. All input signals *above* $-V_{ref}$ fall in a dead zone and are eliminated from the output.

7-5.3 Dead-Zone Circuit with Positive Output

If the diodes in Fig. 7-15 are reversed, the result is a positive-output dead-zone circuit as shown in Fig. 7-16. Reference voltage V_{ref} is found from Eq. (7-1b). $V_{ref} = -15$ V$/3 = -5$ V. Whenever E_i goes above $-V_{ref} = -(-5$ V$) = +5$ V, the output V_{OB} tells by how much E_i exceeds $-V_{ref}$. The dead zone exists for all values of E_i *below* $-V_{ref}$.

7-5.4 Bipolar-Output Dead-Zone Circuit

The positive and negative output dead-zone circuits can be combined as shown in Fig. 7-17 and discussed in Fig. 7-18. The V_{OA} outputs from Figs. 7-15 and 7-16 are connected to an inverting adder. The adder output V_{OB} tells

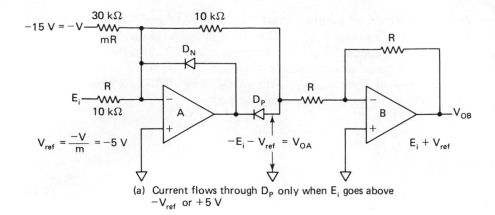

(a) Current flows through D_P only when E_i goes above $-V_{ref}$ or $+5$ V

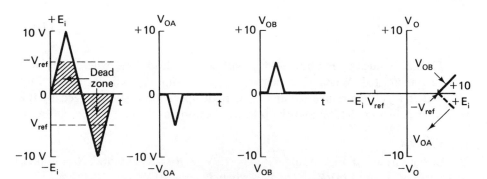

(b) Waveshapes for the positive-output dead-zone circuit

Figure 7-16 Positive-output dead-zone circuit.

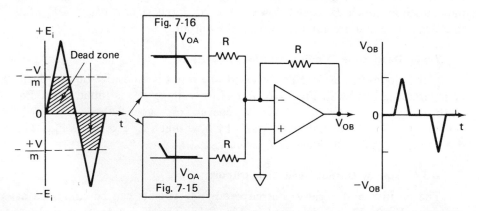

Figure 7-17 The V_{OA} outputs of Figs. 7-15 and 7-16 are combined by an inverting adder to give a bipolar output dead-zone circuit.

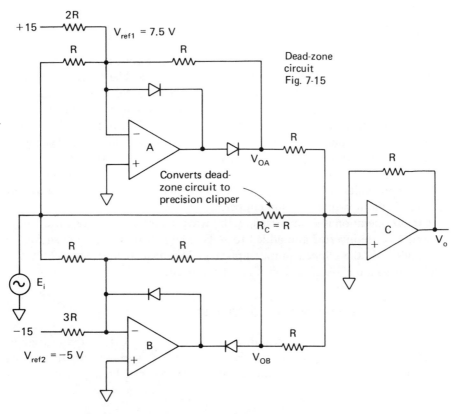

(a) Adding a resistor R_C to the dead-zone circuit of Fig. 7-17 gives a precision clipper

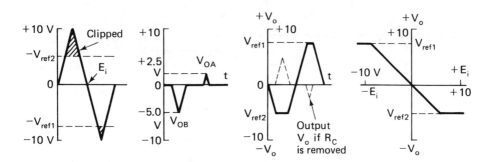

(b) Waveshapes for the precision clipper

Figure 7-18 A precision clipper is made from a dead-zone circuit plus an added resistor R_C.

how much E_i goes above one positive reference voltage and also how much E_i goes below a different negative reference voltage.

7-6 PRECISION CLIPPER

A *clipper* or *amplitude limiter* circuit clips off all signals above a positive reference voltage and all signals below a negative reference voltage. The reference voltages can be made symmetrical or nonsymmetrical around zero. Construction of a precision clipper circuit is accomplished by adding a single resistor, R_C, to a bipolar output dead-zone circuit as shown in Fig. 7-18. The outputs of op amps *A* and *B* are each connected to the input of an inverting adder. Input signal E_i is connected to a third input of the inverting adder, via resistor R_C. If R_C is removed, the circuit would act as a dead-zone circuit. However when R_C is present, input voltage E_i is subtracted from the dead-zone circuit's output and the result is an inverting precison clipper.

   Circuit operation is summarized by the waveshapes in Fig. 7-18(b). Outputs V_{OA} and V_{OB} are inverted and added to $-E_i$. The plot of V_o vs. time shows by solid lines how the clipped output appears. The dashed lines show how the circuit acts as a dead-zone circuit if R_C is removed.

7-7 TRIANGLE-TO-SINE WAVE CONVERTER

Variable frequency sine-wave oscillators are much harder to build than variable-frequency triangle-wave generators. The circuit of Fig. 7-19 converts the output of a triangular-wave generator into a sine wave that can be adjusted for less

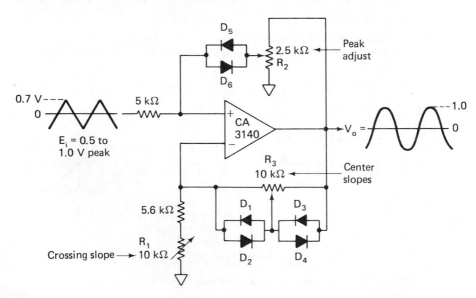

Figure 7-19 Triangle wave-to-sine wave shaper.

than 5% distortion. The triangle–sine converter is an amplifier whose gain varies inversely with amplitude of the output voltage.

R_1 and R_3 set the slope of V_o at low amplitudes near the zero crossings. As V_o increases, the voltage across R_3 increases to begin forward biasing D_1 and D_3 for positive outputs, or D_2 and D_4 for negative outputs. When these diodes conduct, they shunt feedback resistance R_3, lowering the gain. This tends to shape the triangular output above about 0.4 V into a sine wave. In order to get flat tops for the sine-wave output, R_2 and diodes D_5 and D_6 are adjusted to make amplifier gain approach zero at the peaks of V_o.

The circuit is adjusted by comparing a 1 kHz sine wave and the output of the triangle–sine converter on a dual-trace CRO. R_1, R_2, R_3, and the peak amplitude of E_i are adjusted in sequence for best sinusoidal shape. The adjustments interact, so they should be repeated as necessary.

PROBLEMS

7-1. What is the absolute value of $+3$ V and -3 V?

7-2. If the peak value of $E_{in} = 0.5$ V in Fig. 7-1, sketch the waveshapes of V_o vs. t and V_o vs. E_{in} for both a silicon and an ideal diode.

7-3. If E_i is a sine wave with a peak value of 1 V in Figs. 7-2 and 7-3, sketch the waveshapes of V_o vs. t and V_o vs. E_i.

7-4. If diodes D_1 and D_2 are reversed in Fig. 7-2, how does V_o depend on E_i?

7-5. Sketch the circuit for a signal polarity separator.

7-6. Let both diodes be reversed in Fig. 7-8. What is the value of V_o if $E_i = +1$ V or $E_i = -1$ V?

7-7. What is the name of a circuit that follows the voltage peaks of a signal and stores the highest value?

7-8. How do you reset the hold capacitor's voltage to zero in a peak follower-and-hold circuit?

7-9. How do you convert the absolute-value amplifier of Fig. 7-13 to an ac-to-dc converter?

7-10. If resistor mR is changed to 50 kΩ in Example 8-1, find (a) V_{ref}; (b) V_{OA} when $E_i = 10$ V; (c) V_{OB} when $E_i = 10$ V.

7-11. If resistor R_C is removed in Fig. 7-18, sketch V_o vs. E_i.

differential, instrumentation, and bridge amplifiers

ᵗₗ

8

8-0 INTRODUCTION

The most useful amplifier for measurement, instrumentation, or control is the *instrumentation amplifier*. It is designed with several op amps and precision resistors, which make the circuit extremely stable and useful where accuracy is important. There are now many integrated circuits and modular versions available in single packages. Unfortunately, these packages are relatively expensive (from $5 to over $100). But when performance and precision are required, the instrumentation amplifier is well worth the price, because its performance cannot be matched by the average op amp.

An inexpensive first cousin to the instrumentation amplifier is the basic *differential amplifier*. This chapter begins with the differential amplifier to show in which applications it can be superior to the ordinary inverting or noninverting amplifier. The differential amplifier, with some additions, leads into the instrumentation amplifier, which is discussed in the second part of this chapter. The final sections consider *bridge amplifiers*, which involve both instrumentation and basic differential amplifiers.

8-1 BASIC DIFFERENTIAL AMPLIFIER

8-1.1 Introduction

The differential amplifier can measure as well as amplify small signals that are buried in much larger signals. How the differential amplifier accomplishes this task will be studied in Section 8-2. But first, let us build and analyze the circuit performance of the basic differential amplifier.

Four precision (1%) resistors and an op amp make up a differential amplifier, as shown in Fig. 8-1. There are two input terminals, labeled (−) input, and (+) input, corresponding to the closest op amp terminal. If E_1 is replaced by a

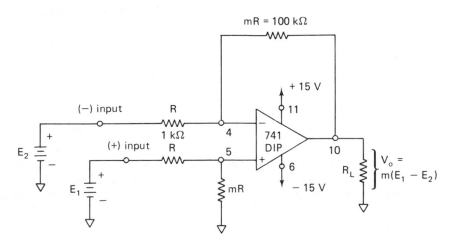

Figure 8-1   Basic differential amplifier.

short circuit, E_2 sees an inverting amplifier with a gain of $-m$. Therefore, the output voltage due to E_2 is $-mE_2$. Now let E_2 be short-circuited; E_1 divides between R and mR to apply a voltage of $E_1 m/(1 + m)$ at the op amp's (+) input. This divided voltage sees a noninverting amplifier with a gain of $(m + 1)$. The output voltage due to E_1 is the divided voltage, $E_1 m/(1 + m)$, times the noninverting amplifier gain, $(1 + m)$, which yields mE_1. Therefore, E_1 is amplified at the output by the multiplier m to mE_1. When both E_1 and E_2 are present at the (+) and (−) inputs, respectively, V_o is $mE_1 - mE_2$, or

$$V_o = mE_1 - mE_2 = m(E_1 - E_2) \tag{8-1}$$

Equation (8-1) shows that the output voltage of the differential amplifier, V_o, is proportional to the *difference* in voltage applied to the (+) and (−) inputs. Multiplier m is called the *differential gain* and is set by the resistor ratios.

Example 8-1

In Fig. 8-1, the differential gain is found from

$$m = \frac{mR}{R} = \frac{100 \text{ k}\Omega}{1 \text{ k}\Omega} = 100$$

Find V_o for $E_1 = 10 \text{ mV}$ and (a) $E_2 = 10 \text{ mV}$, (b) $E_2 = 0 \text{ mV}$, and (c) $E_2 = -20 \text{ mV}$.

Solution. By Eq. (8-1), (a) $V_o = 100(10 - 10)\,\text{mV} = 0$; (b) $V_o = 100(10 - 0)\,\text{mV} = 1.0\,\text{V}$; (c) $V_o = 100[10 - (-20)]\,\text{mV} = 100(30\,\text{mV}) = 3\,\text{V}$.

As expected from Eq. (8-1) and shown from part (a) of Example 8-1, when $E_1 = E_2$ the output voltage is 0. To put it another way, when a common (same) voltage is applied to the input terminals, $V_o = 0$. Section 8-1.2 examines this idea of a common voltage in more detail.

8-1.2 Common-Mode Voltage

The output of the differential amplifier should be 0 when $E_1 = E_2$. The simplest way to apply equal voltages is to wire both inputs together and connect them to the voltage source (see Fig. 8-2). For such a connection, the input voltage is called the *common-mode input voltage*, E_{CM}. Now V_o will be 0 if the resistor ratios are equal (mR to R for the inverting amplifier gain equals mR to R of the voltage-divider network). Practically, the resistor ratios are equalized by installing a potentiometer in series with one resistor, as shown in Fig. 8-2. The potentiometer is trimmed until V_o is reduced to a negligible value. This

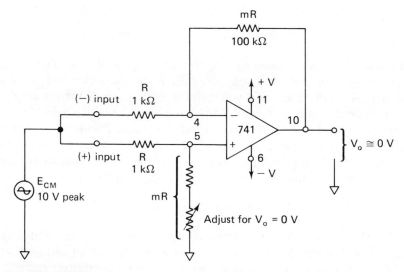

Figure 8-2 The common-mode voltage gain should be zero.

causes the *common-mode voltage gain*, V_o/E_{CM}, to approach 0. It is this characteristic of a differential amplifier that allows a small signal to be picked out of a larger signal. It may be possible to arrange the circuit so that the larger undesired signal is the common-mode input voltage and the small signal is the differential input voltage. Then the differential amplifier's output voltage will contain only an amplified version of the differential input voltage. This possibility is investigated in Section 8-2.

8-2 DIFFERENTIAL VERSUS SINGLE-INPUT AMPLIFIERS

8-2.1 Measuring with a Single-Input Amplifier

A simplified wiring diagram of an inverting amplifier is shown in Fig. 8-3. The power common terminal is shown connected to earth ground. Earth ground comes from a connection to a water pipe on the street side of the water meter. Ground is extended via conduit or a bare Romex wire to the third (green wire) of the instrument line cord and finally to the chassis of the amplifier. This equipment or chassis ground is made to ensure the safety of human operators. It also helps to drain off static charges or any capacitive coupled noise currents to earth.

The signal source is also shown to be earth-grounded in Fig. 8-3. Even if it were not grounded, there would be a leakage resistance or capacitance coupling to earth, to complete the ground loop shown.

Inevitably, noise currents and noise voltages abound from a variety of sources that are often not easily identifiable. The net effect of all this noise is modeled by noise voltage source E_n in Fig. 8-3. It is evident that E_n is in series with signal voltage E_i, so that both are amplified by a factor of -100 due to the inverting amplifier. E_n may be much larger than E_i. For example, the skin signal voltage due to heart beats is less than 1 mV, whereas the body's noise

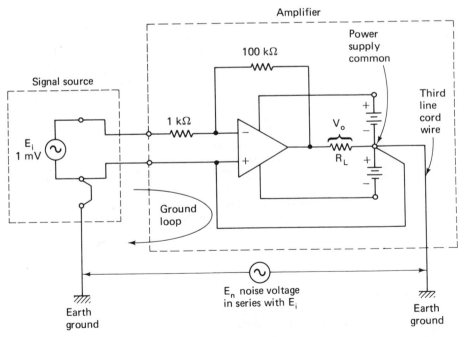

Figure 8-3 Noise voltages act as if they are in series with the input signal of a single input signal. Consequently, both are amplified equally. This arrangement is unworkable if E_n is equal or greater than E_i.

voltage may be tenths of volts or more. So it would be impossible to make an EKG measurement with a single-input amplifier. What is needed is an amplifier that can distinguish between E_i and E_n and amplify only E_i—the differential amplifier.

8-2.2   Measurement with a Differential Amplifier

A differential amplifier is employed to measure only the signal voltage (see Fig. 8-4). The signal voltage E_i is connected across the $(+)$ and $(-)$ inputs of the differential amplifier. Therefore, E_i is amplified by a gain of -100. Noise voltage E_N becomes the common-mode voltage input voltage to the differential amplifier as in Fig. 8-2. Therefore, the noise voltage is *not* amplified and has been effectively eliminated from having any significant effect on the output V_o.

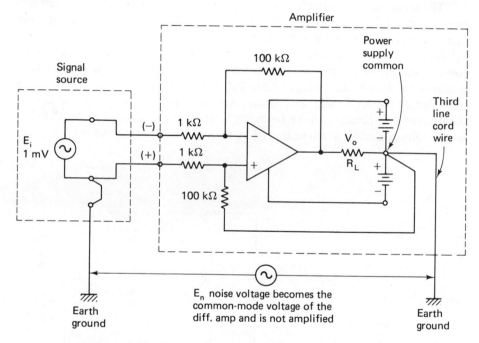

Figure 8-4   The differential amplifier is connected so that noise voltage becomes the common-mode voltage and is *not* amplified. Only the signal voltage E_i is amplified because it has been connected as the differential input voltage.

8-3   IMPROVING THE BASIC DIFFERENTIAL AMPLIFIER

8-3.1   Increasing Input Resistance

There are two disadvantages to the basic differential amplifier studied thus far: It has low input resistance, and changing gain is difficult, because the resistor ratios must be closely matched. The first disadvantage is eliminated by *buffering* or isolating the inputs with voltage followers. This is accomplished with two op amps connected as voltage followers in Fig. 8-5(a). The output of op amp A_1 with respect to ground is E_1, and the output of op amp A_2 with respect to ground is E_2. The differential output voltage V_o is developed across the load resistor R_L. V_o equals the difference between E_1 and E_2 ($V_o = E_1 - E_2$). Note that the output of the basic differential amplifier of Fig. 8-1 is a single-ended output; that is, one side of R_L is connected to ground, and V_o is measured from the output pin of the op amp to ground. The buffered differential amplifier of Fig. 8-5(a) is a differential output; that is, neither side of R_L is connected to ground, and V_o is measured only across R_L.

8-3.2 Adjustable Gain

The second disadvantage of the basic differential amplifier is the lack of adjustable gain. This problem is eliminated by adding three more resistors to the buffered amplifier. The resulting buffered differential-input to differential-output amplifier with adjustable gain is shown in Fig. 8-5(b). The high input resistance is preserved by the voltage followers.

Since the differential input voltage of each op amp is 0 V, the voltages at points 1 and 2 (with respect to ground) are respectively equal to E_1 and E_2. Therefore, the voltage across resistor aR is $E_1 - E_2$. Resistor aR is a potentiometer that is used to adjust the gain. Current through aR is

$$I = \frac{E_1 - E_2}{aR} \tag{8-2}$$

When E_1 is greater than E_2, the direction of I is as shown in Fig. 8-5(b). I flows through both resistors labeled R, and the voltage across all three resistors establishes the value of V_o. In equation form,

$$V_o = (E_1 - E_2)\left(1 + \frac{2}{a}\right) \tag{8-3}$$

where

$$a = \frac{aR}{R}$$

Example 8-2

In Fig. 8-5(b), $E_1 = 10$ mV and $E_2 = 5$ mV. If $aR = 2$ kΩ and $R = 9$ kΩ, find V_o.

Solution. Since $aR = 2$ kΩ and $R = 9$ kΩ,

$$\frac{aR}{R} = \frac{2 \text{ k}\Omega}{9 \text{ k}\Omega} = \frac{2}{9} = a$$

From Eq. (8-3),

$$1 + \frac{2}{a} = 1 + \frac{2}{2/9} = 10$$

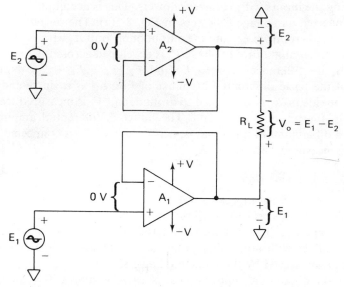

(a) Buffered differential-input to differential-output amplifier

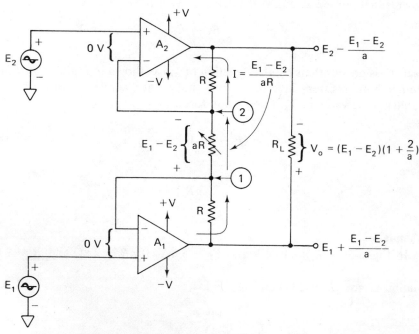

(b) Buffered differential-input to differential-output amplifier with adjustable gain

Figure 8-5 Improving the basic differential amplifier.

Finally,

$$V_o = (10 \text{ mV} - 5 \text{ mV})(10) = 50 \text{ mV}$$

Conclusion

To change the amplifier gain, *only a single resistor aR* now has to be adjusted. However, the buffered differential amplifier has one disadvantage: It can only drive floating loads. *Floating loads* are loads that have neither terminal connected to ground. To drive grounded loads, a circuit must be added that converts a differential input voltage to a single-ended output voltage. Such a circuit is the basic differential amplifier. The resulting circuit configuration, to be studied in Section 8-4, is called an *instrumentation amplifier*.

8-4 INSTRUMENTATION AMPLIFIER

8-4.1 Circuit Operation

The instrumentation amplifier is one of the most useful, precise, and versatile amplifiers available today. It is made from three op amps and seven resistors, as shown in Fig. 8-6. To simplify circuit analysis, note that the instrumentation amplifier is actually made by connecting a buffered amplifier [Fig. 8-5(b)] to a basic differential amplifier (Fig. 8-1). Op amp A_3 and its four equal resistors R form a differential amplifier with a gain of 1. Only the A_3 resistors have to be matched. The primed resistor R' can be made variable to balance out any

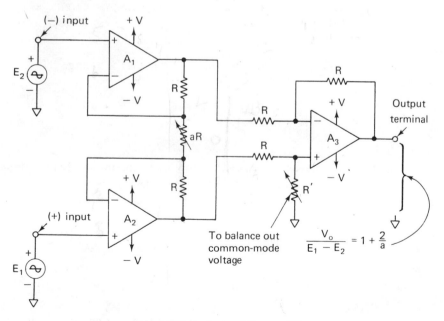

Figure 8-6 Instrumentation amplifier.

common-mode voltage, as shown in Fig. 8-2. Only one resistor, aR, is used to set the gain according to Eq. (8-3), repeated here for convenience:

$$\frac{V_o}{E_1 - E_2} = 1 + \frac{2}{a} \qquad (8\text{-}3)$$

where $a = aR/R$.

E_1 is applied to the $(+)$ input and E_2 to the $(-)$ input. V_o is proportional to the difference between input voltages. Characteristics of the instrumentation amplifier are summarized as follows:

1. The voltage gain, from differential input $(E_1 - E_2)$ to single-ended output, is set by *one* resistor.
2. The input resistance of both inputs is very high and does not change as the gain is varied.
3. V_o does *not* depend on the voltage common to both E_1 and E_2 (common-mode voltage), only on their difference.

Example 8-3

In Fig. 8-6, $R = 25$ kΩ and $aR = 50$ Ω. Calculate the voltage gain.
Solution. From Eq. (8-3),

$$\frac{aR}{R} = \frac{50}{25,000} = \frac{1}{500} = a$$

$$\frac{V_o}{E_1 - E_2} = 1 + \frac{2}{a} = 1 + \frac{2}{1/500} = 1 + (2 \times 500) = 1001$$

Example 8-4

If aR is removed in Fig. 8-6 so that $aR = \infty$, what is the voltage gain?
Solution. $a = \infty$, so

$$\frac{V_o}{E_1 - E_2} = 1 + \frac{2}{\infty} = 1$$

Example 8-5

In Fig. 8-6, the following voltages are applied to the inputs. Each voltage polarity is given with respect to ground. Assuming the gain of 1001 from Example 8-3, find V_o for (a) $E_1 = 5.001$ V and $E_2 = 5.002$ V; (b) $E_1 = 5.001$ V and $E_2 = 5.000$ V; (c) $E_1 = -1.001$ V, $E_2 = 1.002$ V.
Solution. (a)

$$V_o = 1001(E_1 - E_2) = 1001(5.001 - 5.002) \text{ V}$$

$$= 1001(-0.001) \text{ V} = -1.001 \text{ V}$$

(b) $V_o = 1001(5.001 - 5.000) \text{ V} = 1001(0.001) \text{ V} = 1.001$ V
(c) $V_o = 1001[-1.001 - (-1.002)] \text{ V} = 1001(0.001) \text{ V} = 1.001$ V

8-4.2 Referencing Output Voltage

In some applications, it is desirable to offset the output voltage to a reference level other than 0 V. For example, it would be convenient to position a pen on a chart recorder or oscilloscope trace from a control on the instrumentation amplifier. This can be done quite easily by adding a reference voltage in series with one resistor of the basic differential amplifier. Assume that E_1 and E_2 are set equal to 0 V in Fig. 8-6. The outputs of A_1 and A_2 will equal 0 V. Thus we can show the inputs of the differential amplifiers as 0 V in Fig. 8-7.

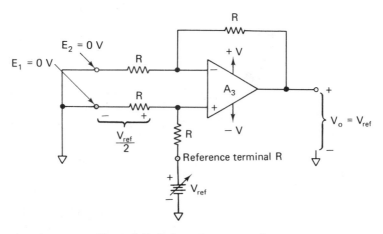

Figure 8-7 Referencing output voltage.

A reference voltage V_{ref} is inserted in series with reference terminal R. V_{ref} is divided by 2 and applied to the A_3 op amp's ($+$) input. Then the non-inverting amplifier gives a gain of 2 so that V_o equals V_{ref}. Now V_o can be set to any desired reference value by adjusting V_{ref}. In practice V_{ref} is the output of a voltage-follower circuit.

8-5 SENSING AND MEASURING WITH THE INSTRUMENTATION AMPLIFIER

8-5.1 Sense Terminal

The versatility and performance of the instrumentation amplifier can be improved by breaking the negative feedback loop around op amp A_3 and bringing out three terminals. As shown in Fig. 8-8, these terminals are *output* terminal O, *sense* terminal S, and *reference* terminal R. If long wires or a current-boost transistor are required between the instrumentation amplifier and load, there will be voltage drops across the connecting wires. To eliminate these voltage drops, the sense terminal and reference terminal are wired directly to the load. Now, wire resistance is added equally to resistors in series with the sense and reference terminals to minimize any mismatch. Even more important, by sensing

163

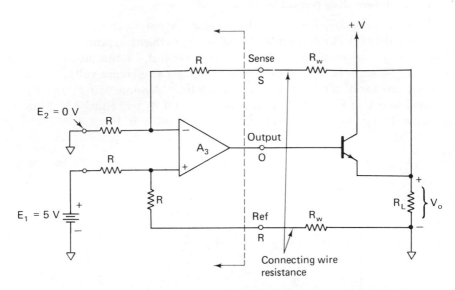

Figure 8-8 Extending the sense and reference terminals to the load makes V_o depend on the amplifier gain and the input voltages, not on the wire resistance.

voltage at the load terminals and *not* at the amplifier's output terminal, feedback acts to hold load voltage constant. If only the basic differential amplifier is used, the output voltage is found from Eq. (8-1) with $m = 1$. If the instrumentation amplifier is used, the output voltage is determined from Eq. (8-3).

This technique is also called *remote voltage sensing;* that is, you sense and control the voltage at a remote load and not at the amplifier's output terminals.

8-5.2 Current and Differential Voltage Measurements

The schematic drawing of an instrumentation amplifier is shown in Fig. 8-9(a). *S* is the sense terminal and *R* is the reference terminal. It is easy to measure the voltage across R_1 with an instrumentation amplifier. Connect the $(+)$ and $(-)$ inputs across R_1 and read V_o with a vacuum-tube voltmeter (VTVM). Then calculate the voltage across R_1 or $E_1 - E_2$ from Eq. (8-3):

$$E_1 - E_2 = \frac{V_o}{1 + 2/a} \tag{8-3}$$

Figure 8-9(a) can be used to determine the current in a circuit. Let R_1 be placed in the circuit and its value be small enough not to disturb circuit operation but large enough to allow current sensing. That is, if we know the value of R_1 and can measure $E_1 - E_2$ as noted above, then the current I can be found from

$$I = \frac{E_1 - E_2}{R_1} \tag{8-4}$$

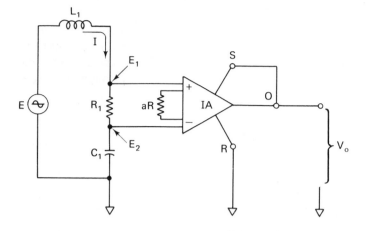

(a) Current measurement with an instrumentation amplifier

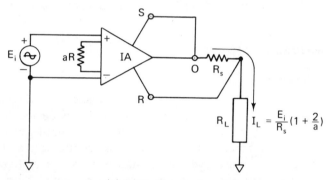

(b) Controlling load current

Figure 8-9   Current measurement and current control using the instrumenta-tion amplifier.

Example 8-6

If $a = \frac{1}{2}$ and $V_o = 10$ V in Fig. 8-9(a), find the differential voltage across resistor R_1.

Solution. By Eq. (8-3),

$$E_1 - E_2 = \frac{V_o}{1 + 2/a} = \frac{10 \text{ V}}{5} = 2 \text{ V}$$

Example 8-7

If $R_1 = 1$ kΩ in Example 8-6, find I.

Solution.

$$I = \frac{V_o}{R_1(1 + 2/a)} = \frac{10 \text{ V}}{1 \text{ kΩ}(5)} = \frac{2 \text{ V}}{1 \text{ kΩ}} = 2 \text{ mA}$$

8-5.3 Controlling Load Current

An instrumentation amplifier can be used to supply a constant current to a load. A current-sensing resistor R_S is also employed in Fig. 8-9(b) to sense the load current. The sense and reference terminals draw negligible current with respect to the load current, I_L. Load current flows through both R_S and R_L and is given by

$$I_L = \frac{E_i}{R_S}\left(1 + \frac{2}{a}\right) \tag{8-5}$$

Example 8-8

In Fig. 8-9(b), $E_i = 1$ mV, $a = \frac{1}{2}$, and $R_S = 5\ \Omega$. Calculate the load current I_L.

Solution. From Eq. (8-5),

$$I_L = \frac{1\ \text{mV}}{5\ \Omega}\left(1 + \frac{2}{1/2}\right) = 1\ \text{mA}$$

E_i controls the load current, and we have a voltage-controlled, *current source*.

8-6 BASIC BRIDGE AMPLIFIER

8-6.1 Introduction

An op amp, three equal resistors, and a transducer form the basic bridge amplifier in Fig. 8-10(a). The transducer in this case is any device that converts an environmental change to a resistance change. For example, a *thermistor* is a transducer whose resistance increases as its temperature decreases. A *photoconductive cell* is a transducer whose resistance decreases as light intensity increases. For circuit analysis, the transducer is represented by a resistor R plus a *change* in resistance ΔR. R is the resistance value at the desired reference, and ΔR is the amount of change in R. For example, a UUA 41J1 thermistor has a resistance of 10,000 Ω at a reference of 25°C. A temperature change of $+1$ to 26°C results in a thermistor resistance of 9573 Ω. ΔR is found to be negative from

$$R_{\text{transducer}} = R_{\text{reference}} + \Delta R$$
$$9573\ \Omega = 10,000\ \Omega + \Delta R$$
$$\Delta R = -427\ \Omega$$

To operate the bridge, we need a stable bridge voltage E, which may be either ac or dc. E should have an internal resistance that is small with respect to R. The simplest way to generate E is to use a voltage divider across the stable supply voltages as shown in Fig. 8-10(b). Then connect a simple voltage follower to the divider. For the resistor values shown, E can be adjusted between $+5$ V and -5 V.

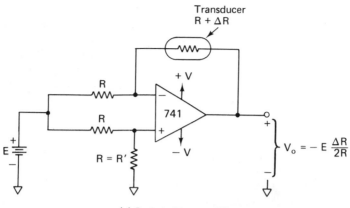

(a) Basic bridge amplifier

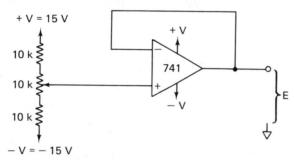

(b) Inexpensive low-resistance voltage source for E

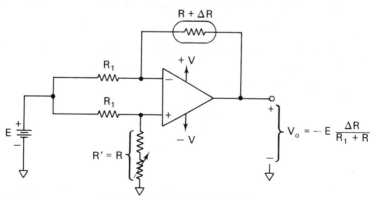

(c) Practical basic bridge amplifier

Figure 8-10 Basic bridge amplifiers and bridge voltage E.

8-6.2 Operation

In Fig. 8-10(a), we assumed that the three matched resistors are equal to the transducer's resistance *at the reference condition*. Unfortunately, this rarely occurs. A more realistic bridge circuit is shown in Fig. 8-10(c). Only two matched resistors, R_1 are required. The primed resistor R' is made up of a series fixed resistor and a variable resistor for the same reasons discussed in Fig. 8-2. The technique of adjusting R' is similar to Fig. 8-2 but must be revised for the bridge circuit as follows.

Zeroing procedure

1. Place the circuit of Fig. 8-10(c) in the reference environment or replace the transducer with a resistance equal to the reference value. For example, if the reference temperature is 25°C, replace the thermistor of Section 8-6.1 with a 10,000-Ω resistor.
2. Adjust R' so that $V_o = 0$ V. R' is now exactly equal to the reference value R.
3. E should be set at the largest value allowable for the application. Typically, $E = 5$ to 15 V. The bridge is now calibrated, and the output voltage V_o will be proportional only to the change in transducer resistance ΔR. V_o may be calculated from

$$V_o = -E\frac{\Delta R}{R_1 + R} \tag{8-6}$$

The minus sign means that V_o goes negative when ΔR is positive.

8-6.3 IC Thermometer

The circuit of Fig. 8-10(c) can be employed as a thermometer. Assuming that the thermistor is a temperature sensor and the circuit is zeroed, a calibrated meter can easily be used to measure temperature, as shown in the next example.

Example 8-9

In the circuit of Fig. 8-10(c), a thermistor transducer is used with the following characteristics:

Resistance (Ω)	Temperature (°C)	ΔR (Ω)
10,450	24	+450
10,000	25	reference = 0
9,573	26	−427

Assume that the circuit has been zeroed at 25°C so that $R = 10,000 \, \Omega$. Find V_o at each temperature if $E_i = 5$ V.

Solution. From Eq. (8-6),

$$\text{At } 24°C \qquad V_o = -5 \text{ V}\left(\frac{450 \, \Omega}{20,000 \, \Omega}\right) = -0.11 \text{ V}$$

At 25°C $\quad V_o = 0 \text{ V}$

At 26°C $\quad V_o = -5 \text{ V} \left(\dfrac{-427 \, \Omega}{20,000 \, \Omega} \right) = 0.10 \text{ V}$

The face of a zero-center voltmeter with a full-scale deflection of ± 0.1 V can be calibrated directly in °C to indicate fractions of a degree between 24 and 26°C.

8-7 ADDING VERSATILITY TO THE BRIDGE AMPLIFIER

8-7.1 Grounded Transducers

In some applications, it may be necessary to have the transducer connected to ground. The standard technique is shown in Fig. 8-11(a). Note that V_o will have the same polarity as E for increases in transducer resistance. The resistor

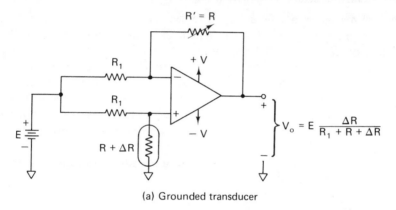

(a) Grounded transducer

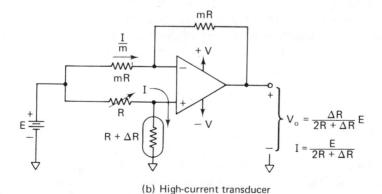

(b) High-current transducer

Figure 8-11 Other ways of using the basic bridge amplifier.

R' is made adjustable and set equal to R of the transducer in accordance with the zeroing procedure stated in Section 8-6.2.

8-7.2   High-Current Transducers

If the current required by the transducer is higher than the current capability of the op amp (5 mA), use the circuit of Fig. 8-11(b). Transducer current is furnished from E. Resistors mR are large enough to hold their currents to about 1 mA, typically 10 kΩ. Transducer current and output voltage may be found from the equations in Fig. 8-11(b). If the transducer current is very small (high-resistance transducers), this same circuit can be used except that the mR resistors will be smaller than R to hold output current of the op amp at about 1 mA.

8-8   THE STRAIN GAGE AND MEASUREMENT OF SMALL RESISTANCE CHANGES

8-8.1   Introduction to the Strain Gage

A strain gage is a conducting wire whose resistance changes by a small amount when it is lengthened or shortened. The change in length is small, a few millionths of an inch. The strain gage is bonded to a structure so that the percent change in length of the strain gage and structure are identical.

A foil-type gage is shown in Fig. 8-12(a). The active length of the gage lies along the transverse axis. The strain gage must be mounted so that its transverse axis lies in the same direction as the structure motion that is to be measured [see Fig. 8-12(b) and (c)]. Lengthening the bar by tension lengthens the strain gage conductor and increases its resistance.

Compression reduces the gage's resistance because the normal length of the strain gage is reduced.

8-8.2   Strain-Gage Material

Strain gages are made from metal alloy such as constantan, Nichrome V, Dynaloy, Stabiloy, or platinum alloy. For high-temperature work they are made of wire. For moderate temperature, strain gages are made by forming the metal alloy into very thin sheets by a photoetching process. The resultant product is called a foil-type strain gage and a typical example is shown in Fig. 8-12(a).

8-8.3   Using Strain-Gage Data

In the next section we will show that our instrumentation measures only the gage's *change* in resistance ΔR. The manufacturer specifies the gage's resistance R. Once ΔR has been measured, the ratio $\Delta R/R$ can be calculated. The manufacturer also furnishes a specified *gage factor* (GF) for each gage. The gage factor is the ratio of the percent change in resistance of a gage to its percent change in length. These percent changes may also be expressed as decimals. If the ratio $\Delta R/R$ is divided by gage factor G, the result is the ratio of the *change* in

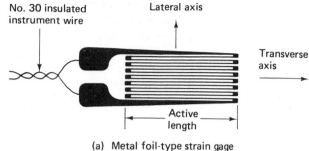

(a) Metal foil-type strain gage

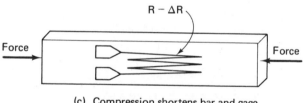

(b) Tension lengthens bar and gage to increase gage resistance by ΔR

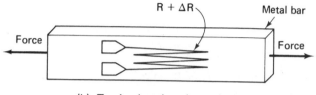

(c) Compression shortens bar and gage to reduce gage resistance by ΔR

Figure 8-12   Using a strain gage to measure the change in length of a structure.

length of the gage ΔL to its original length L. Of course the structure where the gage is mounted has the same $\Delta L/L$. An example will show how gage factor is used.

Example 8-10

A 120-Ω strain gage with a gage factor of 2 is affixed to a metal bar. The bar is stretched and causes a ΔR of 0.001 Ω. Find $\Delta L/L$.

Solution

$$\frac{\Delta L}{L} = \frac{\Delta R/R}{GF} = \frac{0.001 \ \Omega/120 \ \Omega}{2}$$

$$\cong 4.1 \text{ microinches per inch}$$

The ratio $\Delta L/L$ has a name. It is called *unit strain*. It is the unit strain data (we have developed from a measurement of ΔR) that mechanical engineers need. They can use this unit strain data together with known characteristics of the structural material (modules of elasticity) to find the *stress* on the beam. *Stress is the amount of force acting on a unit area.* The unit for stress is pounds per square inch (p.s.i). If the bar in Example 8-10 were made of mild steel, its stress would be about 125 psi. *Strain is the deformation of a material resulting from stress,* or $\Delta L/L$.

8-8.4 Strain-Gage Mounting

Before mounting a strain gage the surface of the mounting beam must be cleaned, sanded, and rinsed with alchohol, Freon, or methyl ethyl ketone (MEK). The gage is then fastened permanently to the cleaned surface by Eastman 910, epoxy, polymide adhesive, or ceramic cement. The manufacturer's procedures should be followed *faithfully*.

8-8.5 Strain-Gage Resistance Changes

It is the *change* of resistance in a strain gage ΔR that must be measured and this change is *small*. ΔR has values of a few milliohms. The technique employed to measure small resistance change is discussed next.

8-9 MEASUREMENT OF SMALL RESISTANCE CHANGES

8-9.1 Need for a Resistance Bridge

To measure resistance we must first find a technique to convert the resistance to a current or voltage for display on an ammeter or voltmeter. If we must measure a small *change* of resistance, we will obtain a very small voltage change. For example, if we passed 5 mA of current through a 120-Ω strain gage, the voltage across the gage would be 0.600 V. If the resistance *changed* by 1 mΩ, the voltage *change* would be 5 μV. To display the 5-μV change, we would need to amplify it by a factor of, for example, 1000 to 5 mV. However, we would also amplify the 0.6 V by 1000 to obtain 600 V plus 5 mV. It is difficult to detect a 5-mV difference in a 600-V signal. Therefore, we need a circuit that allows us to amplify only the *difference* in voltage across the strain gage caused by a *change* in resistance. The solution is found in the bridge circuit.

8-9.2 Basic Resistance Bridge

The strain gage is placed in one arm of a resistance bridge, as shown in Fig. 8-13. Assume that the gage is unstrained, so that its resistance $= R$. Also assume that R_1, R_2, and R_3 are all precisely equal to R. (This unlikely assumption is dealt with in Section 8-10.) Under these conditions $E_1 = E_2 = E/2$ and

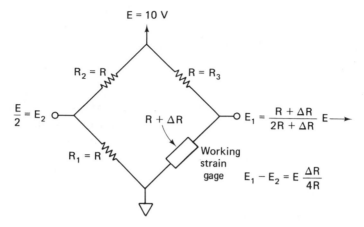

Figure 8-13 The resistor bridge arrangement and supply voltage E convert a resistance change in the strain gage ΔR to a differential output voltage $E_1 - E_2$. If $R = 120\ \Omega$, $E = 10$ V, and $\Delta R = 1$ mΩ, $E_1 - E_2 = 22\ \mu$V.

$E_1 - E_2 = 0$. The bridge is said to be *balanced*. If the strain gage is compressed, R would increase by ΔR and the differential voltage $E_1 - E_2$ would be given by

$$E_1 - E_2 = E\frac{\Delta R}{4R} \qquad (8\text{-}7)$$

Equation (8-7) shows that E should be made large to maximize the bridge differential output voltage, $E_1 - E_2$.

Example 8-11
    If $\Delta R = 0.001\ \Omega$, $R = 120\ \Omega$, and $E = 1.0$ V in Fig. 8-13, find the output of the bridge, $E_1 - E_2$.
Solution. From Eq. (8-7),

$$E_1 - E_2 = 1.0\ \text{V} \times \frac{0.001\ \Omega}{(4)(120)\ \Omega} = 2.2\ \mu\text{V}$$

If E is increased to 10 V, then $E_1 - E_2$ will be increased to 22 μV.
    An instrumentation amplifier can then be used to amplify the differential voltage $E_1 - E_2$ by 1000 to give an output of about 22 mV per milliohm of ΔR.

    We conclude that a voltage E_1 and bridge circuit plus an instrumentation amplifier can convert a change in resistance of 1 mΩ to an output voltage change of 22 mV.

8-9.3 Thermal Effects on Bridge Balance

    Even if you succeed in balancing the bridge circuit of Fig. 8-13, it will not stay in balance because slight temperature changes in the strain gage cause resistance change equal to or greater than those caused by strain. This problem is solved by mounting another identical strain gage immediately adjacent to the working strain gage so that both share the same thermal environment. Therefore, as temperature changes, the added gage's resistance changes exactly as the resistance of the working gage. Thus the added gage provides automatic temperature compensation.

The *temperature-compensation gage* is mounted with its transverse axis perpendicular to the transverse axis of the working gage, as shown in Fig. 8-14. This type of standard gage arrangement is available from manufacturers. The new gage is connected in place of resistor R_1 in the bridge circuit of Fig. 8-13. Once the bridge has been balanced, R of the temperature compensation gage and working gage track one another to hold the bridge in balance. Any unbalance is caused strictly by ΔR of the working gage due to strain.

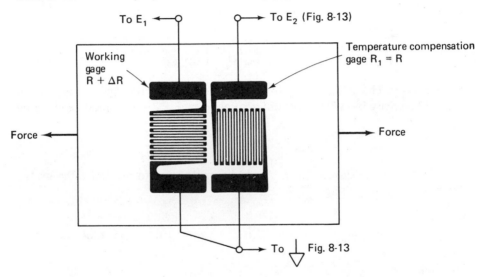

Figure 8-14 The temperature-compensation gage has the same resistance changes as the working gage with changes in temperature. Only the working gage changes resistance with strain. By connecting in the bridge circuit of Fig. 8-13 as shown, resistance changes due to temperature changes are automatically balanced out.

8-10 BALANCING A STRAIN-GAGE BRIDGE

8-10.1 The Obvious Technique

Suppose that you had a working gage and temperature-compensation gage in Fig. 8-15 that are equal to within 1 mΩ. To complete the bridge, you install two 1%, 120-Ω resistors. One is high by 1% at 121.200 Ω and one is low by 1% at 118.800 Ω. They must be equalized to balance the bridge. To do so, a 5-Ω 20-turn balancing pot is installed, as shown in Fig. 8-15. Theoretically, the pot should be set as shown to equalize resistances in the top branches of the bridge at 122.500 Ω.

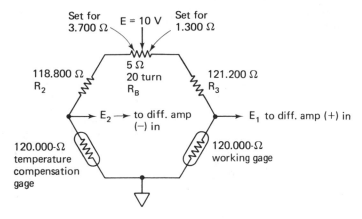

Figure 8-15   Balance pot R_B is adjusted in an attempt to make $E_1 - E_2 = 0$ V.

Further assume that an instrumentation amplifier with a gain of 1000 is connected to the bridge of Fig. 8-15. From Example 8-11, the *output* of the instrumentation amplifier (IA) will be about 22 mV per milliohm of unbalance. This means that the 5-Ω pot must be adjusted to within 1 mΩ of the values shown, so that $E_1 - E_2$ and consequently V_o of the IA will equal 0 V.

Unfortunately, it is very difficult in practice to adjust for balance. This is because each turn of the pot is worth 5 Ω/20 turns = 250 mΩ. When you adjust the pot it is normal to expect a backlash of $\pm\frac{1}{50}$ of a turn. Therefore, your best efforts result in an unbalance at the pot of about ±5 mΩ. You observe this unbalance at the IA's output, where V_o changes by ±0.1 V on either side of zero as you fine-tune the 20-turn pot. It turns out there is a better technique that uses an ordinary linear potentiometer ($\frac{3}{4}$ turn) and a single resistor.

8-10.2 The Better Technique

To analyze operation of the balance network in Fig. 8-16, assume that the R_2 and R_3 bridge resistors are reasonably equal, to within $\pm1\%$. The strain gage's resistance should have equal resistances within several milliohms if the working gage is not under strain.

Resistor R_{B1} is an ordinary $\frac{3}{4}$-turn linear pot. Its resistance should be about $\frac{1}{10}$ or less than resistor R_{B2} so that the voltage fE depends only on E and the decimal fraction f. Values of f vary from 0 to 1.0 as the pot is adjusted from one limit to the other.

Resistor R_{B2} is chosen to be greater than 10 or more times the individual gage resistance, and also 10 or more times R_{B1}. Under these conditions R_{B2} does not load down the voltage-divider action of R_{B1}. Also, the size of R_{B2} determines the maximum balancing current that can be injected into, or extracted from, the E_2 node. The pot setting f determines how much of that maximum current is injected or extracted.

Balancing action is summarized by observing that if $f > 0.5$, a small current,

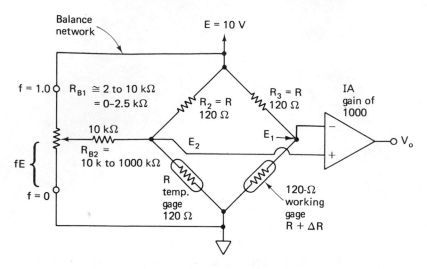

Figure 8-16 Improved balance networks R_{B1} and R_{B2} allow easy adjustment of V_0 to 0 V.

is injected into the E_2 node and flows through the temperature gage to ground, this makes E_2 more positive. If $f < 0.5$ current is extracted from the E_2 node, this increases current through R_2 to make E_2 less positive.

   In a real bridge setup, begin with $R_{B2} = 10$ kΩ and $R_{B1} = 1$ kΩ. Monitor V_o of the IA and check the balancing action. If the variation in V_o is larger than you want, increase R_{B2} to 100 kΩ and recheck the balance action. The final value of R_{B2} is selected by experiment and depends on the magnitude of unbalance between R_2 and R_3.

8-11   INCREASING STRAIN-GAGE BRIDGE OUTPUT

A single working gage and temperature-compensation gage were shown to give a differential bridge output in Fig. 8-13(a) of

$$E_1 - E_2 = E\frac{\Delta R}{4R}$$

This bridge circuit and placement of the gages is shown again in Fig. 8-17(a).

   The bridge output voltage $E_1 - E_2$ can be doubled by doubling the number of working gages, as in Fig. 8-17(b). Bridges 1–2 and 5–6 are the working gages and will increase resistance (tension) if force is applied as shown. By arranging the working gages in opposite arms of the bridge and the temperature gages in the other arms, the bridge output is

$$E_1 - E_2 = E\frac{\Delta R}{2R}$$

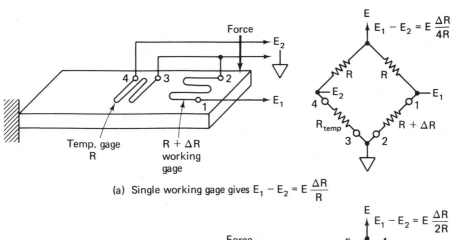

(a) Single working gage gives $E_1 - E_2 = E \dfrac{\Delta R}{R}$

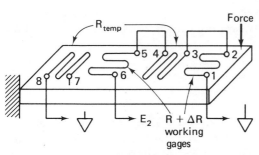

(b) Two working gages double the $E_1 - E_2$ output over that of (a)

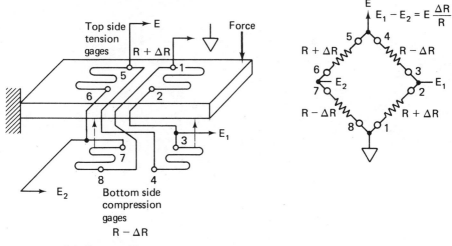

(c) Four working gages quadruple the $E_1 - E_2$ over that of (a)

Figure 8-17 Comparison of sensitivity for three strain-gage bridge arrangements.

177

If the structural member experiences bending as shown in Fig. 8-17(c), even greater bridge sensitivity can be obtained. The upper side of the bar will lengthen (tension) to increase resistance of the working strain gages by $(+)\Delta R$. The lower side of the bar will shorten (compression) to decrease the working strain gages by $(-)\Delta R$.

The compression gages 1–2 and 5–6 are connected in opposite arms of the bridge. Tension gages 3–4 and 7–8 are connected in the remaining opposite arms of the bridge. The gages also temperature-compensate one another. The output of the four-strain-gage arrangement in Fig. 8-17(c) is quadrupled over the single-gage bridge to

$$E_1 - E_2 = E\frac{\Delta R}{R}$$

Of course, each bridge arrangement in Fig. 8-17 should be connected to a balance network (which, for clarity, was not shown; see Fig. 8-16 and Section 8-12).

8-12 A PRACTICAL STRAIN-GAGE APPLICATION

As shown in Fig. 8-18, an AD521 (Analog Devices) instrumentation amplifier (IA) is connected to a bridge arrangement of four strain gages. The gages are 120-Ω, SR4, foil-type strain gages. They are mounted on a steel bar in accordance with Fig. 8-17(c). Also the balance network of Fig. 8-16 is connected to

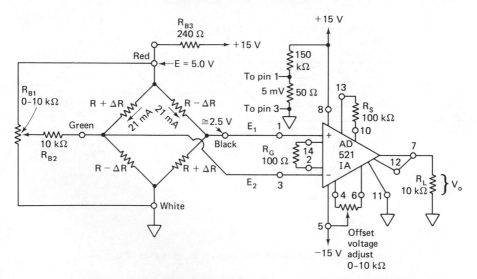

Figure 8-18 The AD521 instrumentation amplifier is used to amplify output of the four working strain gages [see Fig. 8-17(c)].

the strain gage bridge. R_{B2} was selected, after experiment, as 10 kΩ. Strain gages were mounted in strict accordance with the manufacturer's instructions (BLH Electronics, Inc.)

The calibration procedure for the instrumentation is as follows:

1. *Gain resistors:* Select gain setting resistors R_{scale} (R_S) and R_{gain} (R_G) to give a gain of 1000. Gain is set by the ratio of R_S/R_G, or 100 kΩ/100 Ω.
2. *Offset voltage adjust:* With R_S and R_G installed, ground inputs 1 and 3. Adjust the IA's offset voltage pot for $V_o = 0$ V (see Chapter 9).
3. *Gain measurement:* Connect the 5-mV signal from the 150-kΩ and 50-Ω voltage divider as $E_1 - E_2$ to pins 1 and 2. Measure V_o; calculate gain $= V_o/(E_1 - E_2)$.
4. *Zeroing the bridge network:* Connect E_1 and E_2 to pins 1 and 3 of the IA. Adjust R_{B1} for $V_o = 0$ V when the gages are *not* under strain.

Example 8-12

The test setup of Fig. 8-18 is used to measure the strain resulting from deflection of a steel bar. V_o is measured to be 100 mV. Calculate (a) ΔR; (b) $\Delta R/R$; (c) $\Delta L/L$. Assume that the calibration procedure has been followed and that the gain from step 3 is 1000. The gage factor is 2.0.

Solution. (a) Find $E_1 - E_2$ from

$$E_1 - E_2 = \frac{V_o}{\text{gain}} = \frac{100 \text{ mV}}{1000} = 0.1 \text{ mV}$$

Find ΔR from Fig. 8-17(c):

$$\Delta R = \frac{R(E_1 - E_2)}{E} = \frac{120 \ \Omega(0.1 \times 10^{-3} \text{ V})}{5.0 \text{ V}}$$

$$= 0.0024 \ \Omega = 2.4 \text{ m}\Omega$$

(b)

$$\frac{\Delta R}{R} = \frac{0.0024 \ \Omega}{120 \ \Omega} = 0.000020 = 20 \times 10^{-6} \ \mu\Omega/\Omega$$

(c) From gage factor $= (\Delta R/R)/(\Delta L/L)$, we obtain

$$\frac{\Delta L}{L} = \frac{20 \times 10^{-6}}{2} = 10 \times 10^{-6} = 10 \ \mu\text{in./in.}$$

Note: Resistor R_{B3} is selected to restrict gage current *below* 25 *mA* to limit self-heating.

Since we now know the value of strain from $\Delta L/L$, we can look up the *modulus of elasticity* for steel, $E = 30 \times 10^6$. Then *stress* can be calculated from

$$\text{stress} = E \times \text{strain} = (30 \times 10^6)(10 \times 10^{-6})$$
$$= 300 \text{ psi}$$

PROBLEMS

8-1. In Fig. 8-1, $m = 20$, $E_1 = 0.2$ V, and $E_2 = 0.25$ V. Find V_o.

8-2. If $V_o = 10$ V in Fig. 8-1, $E_1 = 7.5$ V, and $E_2 = 7.4$ V, find m.

8-3. If $E_{\text{cm}} = 5.0$ V in Fig. 8-2, find V_o.

8-4. In Fig. 8-3, $E_i = 2$ mV and $E_N = 50$ mV. What is the output voltage due to (a) E_i; (b) E_N?

8-5. In Fig. 8-4, $E_i = 2$ mV and $E_N = 50$ mV. What is the output voltage due to (a) E_i; (b) E_N?

8-6. What is the main advantage of a differential amplifier over an inverting amplifier with respect to an input noise signal voltage?

8-7. Find V_o in Fig. 8-5(b) if $E_1 = -5$ V and $E_2 = -3$ V.

8-8. In Fig. 8-5(b), $R = 10$ kΩ and $aR = 2$ kΩ. If $E_1 = 1.5$ V and $E_2 = 0.5$ V, find V_o.

8-9. In Fig. 8-6 the overall gain is 21 and $V_o = 3$ V. Determine (a) $E_1 - E_2$; (b) a.

8-10. In Fig. 8-6, $R = 25$ kΩ, $aR = 100$ Ω, $E_1 = 1.01$ V, and $E_2 = 1.02$ V. Find V_o.

8-11. If $V_{\text{ref}} = 5.0$ V in Fig. 8-7, find (a) V_o; (b) the voltage at the (+) input with respect to ground.

8-12. R_S is changed to 10 Ω in Example 8-8. Find the new load current I_L.

8-13. E is changed to 10 V in Example 8-9. Calculate V_o for thermistor temperatures of (a) 24°C; (b) 25°C; (c) 26°C.

bias, offsets,
and drift

9-0   INTRODUCTION

The op amp is widely used in amplifier circuits to amplify dc or ac signals or combinations of them. In dc amplifier applications, certain electrical character- istics of the op amp can cause large errors in the output voltage. The ideal output voltage should be equal to the product of the dc input signal and the amplifier's closed-loop voltage gain. However, the output voltage may have an added error component. This error is due to differences between an ideal op amp and a real op amp. If the ideal value of output voltage is large with respect to the error component, then we can ignore the op amp characteristic that causes it. But if the error component is comparable to or even larger than the ideal value, we must try to minimize the error. Op amp characteristics that add error compo- nents to the dc output voltage are:

1. Input bias currents
2. Input offset current
3. Input offset voltage
4. Drift

When the op amp is used in an ac amplifier, coupling capacitors eliminate dc output-voltage error. Therefore, characteristics 1 to 4 above are usually unimportant in ac applications. However, there are new problems for ac amplifiers. They are:

5. Frequency response
6. Slew rate

Frequency response refers to how voltage gain varies as frequency changes. The most convenient way to display such data is by a plot of voltage gain versus frequency. Op amp manufacturers give such a plot for open-loop gain versus frequency. A glance at the plot quickly shows how much gain is obtainable at a particular frequency.

If the op amp has sufficient gain at a particular frequency, there is still a possibility of an error being introduced in V_o. This is because there is a fundamental limit imposed by the op amp (and certain circuit capacitors) on how fast the output voltage can change. If the input signal tells the op amp output to change faster than it can, distortion is introduced in the output voltage. The op amp characteristic responsible for this type of error is its capacitance. This type of error is called *slew-rate limiting*.

Op amp characteristics and the circuit applications that each type of error *may* affect are summarized in Fig. 9-1. The first four characteristics can limit dc performance; the last two can limit ac performance.

Op-amp characteristic that may affect performance	Op-amp application			
	DC amplifier		AC amplifier	
	Small output	Large output	Small output	Large output
1. Input bias current	yes	maybe	no	no
2. Offset current	yes	maybe	no	no
3. Input offset voltage	yes	maybe	no	no
4. Drift	yes	no	no	no
5. Frequency response	no	no	yes	yes
6. Slew rate	no	yes	no	yes

Figure 9-1 Listing of op amp applications and characteristics that affect operation.

Op amp characteristics that cause errors primarily in dc performance are studied in this chapter. Those that cause errors in ac performance are studied in Chapter 10. We begin with input bias currents and ways in which they cause errors in the dc output voltage of an op amp circuit.

9-1 INPUT BIAS CURRENTS

Transistors within the op amp must be *biased* correctly before any signal voltage is applied. Biasing correctly means that the transistor has the right value of base and collector current as well as collector-to-emitter voltage. Until now, we have considered that the input terminals of the op amp conduct no current. This is the ideal condition. Practically, however, the input terminals do conduct a small value of dc current to bias the op amps' transistors (see Appendices 1 and 2). A simplified diagram of the op amp is shown in Fig. 9-2(a). To discuss the effect of input bias currents, it is convenient to model them as current sources in series with each input terminal, as shown in Fig. 9-2(b).

The (−) input's bias current, I_{B-}, will usually not be equal to the (+) input's bias current, I_{B+}. Manufacturers specify an *average* input bias current I_B, which is found by adding the *magnitudes* of I_{B+} and I_{B-} and dividing this sum by 2. In equation form,

$$I_B = \frac{|I_{B+}| + |I_{B-}|}{2} \tag{9-1}$$

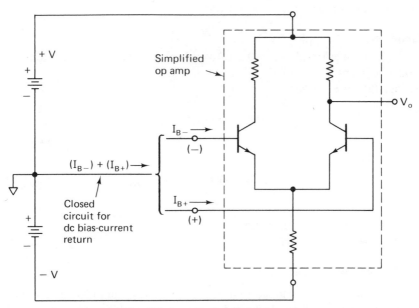

(a) Simplified op amp input circuit

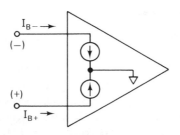

(b) Model for bias currents

Figure 9-2 Origin and model of dc input bias currents.

where $|I_{\text{B}+}|$ is the magnitude of $I_{\text{B}+}$ and $|I_{\text{B}-}|$ is the magnitude of $I_{\text{B}-}$. The range of I_{B} is from 1 μA or more for general-purpose op amps to 1 pA or less for op amps that have field-effect transistors at the input.

9-2 INPUT OFFSET CURRENT

The difference in magnitudes between $I_{\text{B}+}$ and $I_{\text{B}-}$ is called the *input offset current* I_{os}:

$$I_{os} = |I_{\text{B}+}| - |I_{\text{B}-}| \tag{9-2}$$

Manufacturers specify I_{os} for a circuit condition where the output is at 0 V and the temperature is 25°C. The typical I_{os} is less than 25% of I_{B}, for the average input bias current (see Appendices 1 and 2).

Example 9-1
   If $I_{\text{B}+} = 0.4$ μA and $I_{\text{B}-} = 0.3$ μA, find (a) the average bias current I_{B}; (b) the offset current I_{os}.
Solution. (a) By Eq. (9-1),

$$I_{\text{B}} = \frac{(0.4 + 0.3)\mu A}{2} = 0.35 \ \mu A$$

(b) By Eq. (9-2),

$$I_{os} = (0.4 - 0.3) \ \mu A = 0.1 \ \mu A$$

9-3 EFFECT OF BIAS CURRENTS
ON OUTPUT VOLTAGE

9-3.1 Simplification

   In this section it is assumed that bias currents are the only op amp characteristic that will cause an undesired component in the output voltage. The effects of other op amp characteristics on V_o will be dealt with individually.

9-3.2 Effect of (−) Input Bias Current

   Output voltage should ideally equal 0 V in each circuit of Fig. 9-3, because input voltage E_i is 0 V. The fact that a voltage component will be measured is due strictly to I_{B-}. (Assume for simplicity that V_{io}, input offset voltage, is zero. V_{io} is discussed in Section 9-5.) In Fig. 9-3(a), the bias current is furnished from the output terminal. Since negative feedback forces the differential input voltage to 0 V, V_o must rise to supply the voltage drop across R_f. Thus, the output voltage error due to I_{B-} is found from $V_o = R_f I_{B-}$. I_{B+} flows through 0 Ω, so it causes no voltage error.

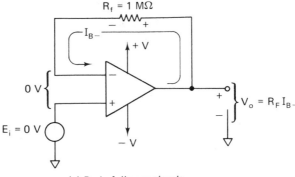

(a) Basic follower circuit

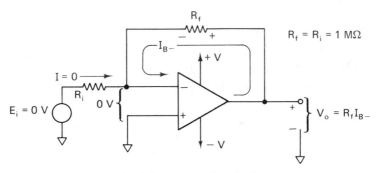

(b) Basic inverting circuit

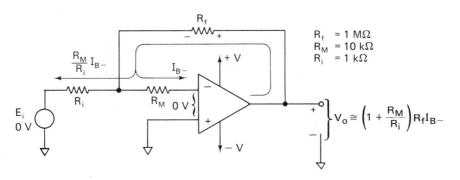

(c) Multiplier resistor R_M increases effect of I_{B-} on V_o

Figure 9-3   Effects of $(-)$ input bias current on output voltage.

The circuit of Fig. 9-3(b) has the same output-voltage error expression, $V_o = R_f I_{B-}$. No current flows through R_i, because there is 0 V on each side of R_i. Thus all of I_{B-} flows through R_f. [Recall that an ideal amplifier with negative feedback has 0 voltage between the $(+)$ and $(-)$ inputs.]

Example 9-2

In Fig. 9-3(a), $V_o = 0.4$ V. Find I_{B-}.
Solution

$$I_{B-} = \frac{V_o}{R_f} = \frac{0.4 \text{ V}}{1 \text{ M}\Omega} = 0.4 \text{ }\mu\text{A}$$

Placing a multiplying resistor R_M in series with the $(-)$ input in Fig. 9-3(c) multiplies the effect of I_{B-} on V_o. I_{B-} sets up a voltage drop across R_M that establishes an equal drop across R_i.

Now both the R_i current and I_{B-} must be furnished through R_f. Thus the error in V_o will be much larger. R_M would be undesirable in a normal circuit; however, we want to measure low values of the bias current, and Fig. 9-3(c) is a way of doing it. For the resistor values shown, $V_o \cong 11 R_f I_{B-}$.

9-3.3   Effect of (+) Input Bias Current

Since $E_i = 0$ V in Fig. 9-4, V_o should ideally equal 0 V. However, the positive input bias current I_{B+} flows through the internal resistance of the signal generator. Internal generator resistance is modeled by resistor R_G in Fig. 9-4.

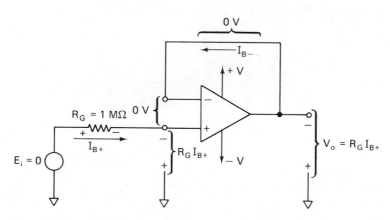

Figure 9-4   Effect of (+) input bias current on output voltage.

I_{B+} sets up a voltage drop of $R_G I_{B+}$ across R_G and applies it to the $(+)$ input. The differential input voltage is 0 V, so the $(-)$ input is also at $R_G I_{B+}$ in Fig. 9-4. Since there is 0 resistance in the feedback loop, V_o equals $R_G I_{B+}$. (The return path for I_{B+} is through $-V$ supply and back to ground.)

Example 9-3

In Fig. 9-4, $V_o = -0.3$ V. Find I_{B+}.
Solution

$$I_{B+} = -\frac{V_o}{R_G} = -\frac{0.3 \text{ V}}{1 \text{ M}\Omega} = -0.3 \text{ }\mu\text{A}$$

9-4 EFFECT OF OFFSET CURRENT ON OUTPUT VOLTAGE

9-4.1 Current-Compensating the Voltage Follower

If I_{B+} and I_{B-} were always equal, it would be possible to compensate for their effects on V_o. For example, in the voltage follower of Fig. 9-5(a), I_{B+} flows through the signal generator resistance R_G. If we insert $R_f = R_G$ in the feedback loop, I_{B-} will develop a voltage drop across R_f of $R_f I_{B-}$. If $R_f = R_G$ and $I_{B+} = I_{B-}$, their voltage drops will cancel each other and V_o will equal 0 V when $E_i = 0$ V. Unfortunately, I_{B+} is seldom equal to I_{B-}. V_o will then be equal to R_G times the difference between I_{B+} and I_{B-} ($I_{B+} - I_{B-} = I_{os}$). Therefore, by making $R_f = R_G$, we have reduced the error in V_o from $R_G I_{B+}$ in Fig. 9-4 to $R_G I_{os}$ in Fig. 9-5(a). Recall that I_{os} is typically 25% of I_B. If the value of I_{os} is too large, an op amp with a smaller value of I_{os} is needed.

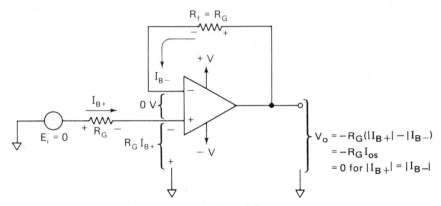

(a) Compensated voltage follower

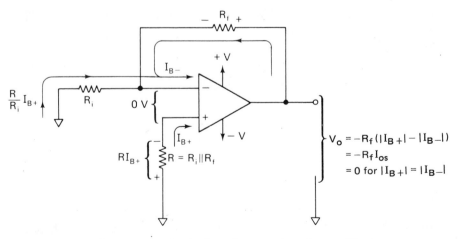

(b) Compensation for inverting or noninverting amplifiers

Figure 9-5 Balancing-out effects of bias current in V_o.

9-4.2 Current-Compensating Other Amplifiers

To minimize errors in V_o due to bias currents for either inverting or non-inverting amplifiers, resistor R as shown in Fig. 9-5(b), must be added to the circuit. With no input signal applied, V_o depends on R_f times I_{os} [where I_{os} is given by Eq. (9-2)]. Resistor R is called the *current-compensating resistor* and is equal to the parallel combination of R_i and R_f, or

$$R = R_i \| R_f = \frac{R_i R_f}{R_i + R_f} \tag{9-3}$$

R_i and R should include any signal generator resistance. By inserting resistor R, the error voltage in V_0 will be reduced more than 25% from $R_f I_{B-}$ in Fig. 9-3(b) to $R_f I_{os}$ in Fig. 9-5(b). In the event that $I_{B-} = I_{B+}$, then $I_{os} = 0$ and $V_o = 0$.

9-4.3 Summary on Bias-Current Compensation

Always add a bias-current compensating resistor R in series with the (+) input terminal (except for FET input op amps). The value of R should equal the parallel combination of all resistance branches connected to the $(-)$ terminal. Any internal resistance in the signal source should also be included in the calculations.

In circuits where more than a single resistor is connected to the (+) input, bias-current compensation is accomplished by observing the following principle. *The dc resistance seen from the (+) input to ground should equal the dc resistance seen from the (−) input to ground.* In applying this principle, signal sources are replaced by their internal dc resistance and the op amp output terminal is considered to be at ground potential.

Example 9-4

(a) In Fig. 9-5(b), $R_f = 100 \text{ k}\Omega$ and $R_i = 10 \text{ k}\Omega$. Find R. (b) If $R_f = 100 \text{ k}\Omega$ and $R_i = 100 \text{ k}\Omega$, find R.
Solution. (a) By Eq. (9-3),

$$R = \frac{(100 \text{ k}\Omega)(10 \text{ k}\Omega)}{100 \text{ k}\Omega + 10 \text{ k}\Omega} = 9.1 \text{ k}\Omega$$

(b) By Eq. (9-3),

$$R = \frac{(100 \text{ k}\Omega)(100 \text{ k}\Omega)}{100 \text{ k}\Omega + 100 \text{ k}\Omega} = 50 \text{ k}\Omega$$

9-5 INPUT OFFSET VOLTAGE

9-5.1 Definition and Model

In Fig. 9-6(a), the output voltage V_o should equal 0 V. However, there will be a small error-voltage component present in V_o. Its value can range from microvolts to millivolts and is caused by very small but unavoidable unbalances

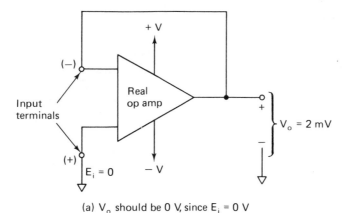

(a) V_o should be 0 V, since E_i = 0 V

(b) Error in V_o is modeled by dc voltage V_{io}
in series with (+) input

Figure 9-6 Effect of input offset voltage in the real op amp of (a) is modeled
by an ideal op amp plus battery V_{io} in (b).

inside the op amp. The easiest way to study the *net effect* of all these internal
unbalances is to visualize a small dc voltage in *series* with one of the input
terminals. This dc voltage is modeled by a battery in Fig. 9-6(b) and is called
input offset voltage, V_{io} (see Appendices 1 and 2 for typical values). Note that
V_{io} is shown in series with the (+) input terminal of the op amp. It makes no
difference whether V_{io} is modeled in series with the (−) input or the (+) input.
But it is easier to determine the polarity of V_{io} if it is placed in series with the
(+) input. For example, if the output terminal is positive (with respect to ground)
in Fig. 9-6(b), V_{io} should be drawn with its (+) battery terminal connected to
the ideal op amp's (+) input.

9-5.2 Effect of Input Offset Voltage
on Output Voltage

Fig. 9-7(a) shows that V_{io} and the large value of the open-loop gain of the op amp act to drive V_o to negative saturation. Contrast the polarity of V_{io} in Figs. 9-6(b) and 9-7(a). If you buy several op amps and plug them into the test circuit of Fig. 9-7(a), some will drive V_o to $+V_{sat}$ and the remainder will drive V_o to $-V_{sat}$. Therefore, the magnitude and polarity of V_{io} varies from op amp to op amp. We also conclude that this is no way to measure V_{io}. To learn how V_{io} affects amplifiers with negative feedback, we study how to measure V_{io}.

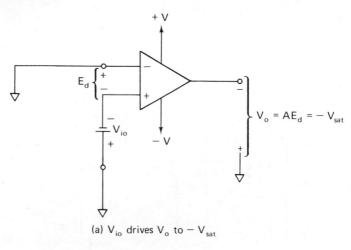

(a) V_{io} drives V_o to $-V_{sat}$

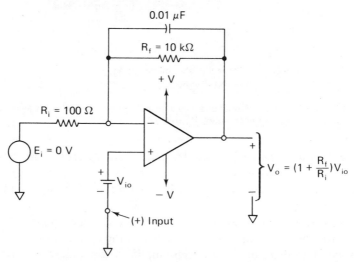

(b) V_{io} is amplified, causing a large error in V_o

Figure 9-7 V_o should be 0 V in (a) and (b) but contains a dc error voltage due to V_{io}. (Error component due to bias current is neglected.)

9-5.3 Measurement of Input Offset Voltage

For simplicity, the effects of bias currents are neglected in the following discussion. Figure 9-7(b) shows how to measure V_{io}. It also shows how to predict the magnitude of error that V_{io} will cause in the output voltage. Since $E_i = 0$ V, V_o should equal 0 V. But V_{io} acts exactly as would a signal in series with the noninverting input. Therefore, V_{io} is amplified exactly as any signal applied to the (+) input of a noninverting amplifier (see Section 3-6). The error in V_o due to V_{io} is given by:

$$\text{error voltage due to } V_{io} = V_{io}\left(1 + \frac{R_f}{R_i}\right) \qquad (9\text{-}4)$$

The output error voltage in Fig. 9-7(b) is given by Eq. (9-4) whether the circuit is used as an inverting or as a noninverting amplifier. That is, E_i could be inserted in series with R_i (inverting amplifier) for a gain of $-(R_f/R_i)$ or in series with the (+) input (noninverting amplifier) for a gain of $1 + (R_f/R_i)$. A bias-current compensating resistor (a resistor in series with the (+) input) has no effect on this type of error.

Conclusion

To measure V_{io}, set up the circuit of Fig. 9-7(b). The capacitor is installed across R_f to minimize noise in V_o. Measure V_o, R_f, and R_i. Calculate V_{io} from

$$V_{io} = \frac{V_o}{1 + R_f/R_i} \qquad (9\text{-}5)$$

Note that R_f is made small to minimize the effect of input bias current.

Example 9-5

V_{io} is specified to be 1 mV for a 741-type op amp. Predict the value of V_o that would be measured in Fig. 9-7(b).

Solution. From Eq. (9-5),

$$V_o = \left(1 + \frac{10,000}{100}\right)(1 \text{ mV}) = 101 \text{ mV}$$

9-6 INPUT OFFSET VOLTAGE FOR THE ADDER CIRCUIT

9-6.1 Comparison of Signal Gain and Offset Voltage Gain

In both inverting and noninverting amplifier applications, the input offset voltage V_{io} is multiplied by $(1 + R_f/R_i)$. The input signal in either circuit is multiplied by a different gain. R_f/R_i is the gain for the inverter and $(1 + R_f/R_i)$ for the noninverter. In the inverting adder circuit of Fig. 9-8(a) (neglecting bias currents), V_{io} is multiplied by a larger number than the signal at each input.

For example, in Fig. 9-8(a) signals E_1 and E_2 are each larger than V_{io} but E_1 is multiplied by $-R_f/R_1 = -1$ and develops a component of -5 mV in V_o. E_2 is likewise multiplied by -1 and adds a -5-mV component to V_o. Thus the correct value of V_o should be -10 mV. Since E_3 is 0 its contribution to V_o is 0 (see Section 3-2).

If we temporarily let E_1 and $E_2 = 0$ V in Fig. 9-8(a), the $(-)$ input sees three equal resistors forming parallel paths to ground. The single equivalent

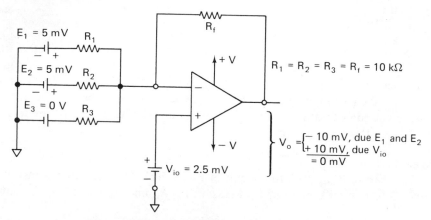

(a) V_o has a -10-mV component due to
E_1 and E_2 plus a 10-mV error component
due to V_{io}

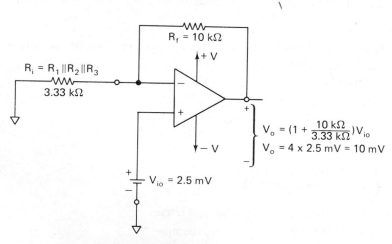

(b) V_{io} is multiplied by a gain of 4 to generate
a 10-mV error component in V_o

Figure 9-8 Each input voltage of the inverting adder in (a) is multiplied
by a gain of -1. V_{io} is multiplied by a gain of $+4$.

series resistance, R_i, is shown in Fig. 9-8(b). For three equal 10-kΩ resistors in parallel, the equivalent resistance R_i is found by $10\,k\Omega/3 = 3.33\,k\Omega$. V_{io} is amplified just as in Fig. 9-7(b) to give an output error of $+10\,mV$. Therefore, the total output voltage in Fig. 9-8(a) is 0 instead of $-10\,mV$.

Conclusions

In an adder circuit, the input offset voltage has a gain of *1 plus the number of inputs*. The more inputs, the greater the error component in the output voltage. Since the gain for the inputs is -1, *the offset voltage gain always exceeds the signal voltage gain*.

9-6.2 How Not to Eliminate the Effects of Offset Voltage

One might be tempted to add an input to the adder such as E_3 in Fig. 9-8(a) to balance out the effect of V_{io}. For example, if E_3 is made equal to $10\,mV$, then E_3, R_3, and R_f will add a $-10\,mV$ component to V_o and balance out the $+10\,mV$ due to V_{io}. There are two disadvantages to this approach. First, such a small value of E_3 would have to be obtained from a resistor-divider network between the power supply terminals of $+V$ and $-V$. The second disadvantage is that any resistance added between the $(-)$ input and ground raises the *noise gain*. This situation is treated in Sections 10-5.3 and 10-5.4.

Section 9-7 shows how to minimize the output voltage errors caused by both bias currents and input offset voltage.

9-7 NULLING-OUT EFFECTS OF OFFSET VOLTAGE AND BIAS CURRENTS

9-7.1 Design or Analysis Sequence

To minimize dc error voltages in the output voltage, follow this sequence:

1. Select a bias-current compensating resistor in accordance with the principles set forth in Section 9-4.3.
2. Get a circuit for minimizing effects of the input offset voltage from the manufacturer's data sheet. This principle is treated in more detail in Section 9-7.2 and in Appendices 1 and 2.
3. Go through the output-voltage nulling procedure given in Section 9-7.3.

9-7.2 Null Circuits for Offset Voltage

It is possible to imagine a fairly complex resistor-divider network that would inject a small variable voltage into the $(+)$ or $(-)$ input terminal. This would compensate for the effects of both input offset voltage and offset current. However, the extra components are more costly and bulky than necessary. It is far better to go to the op amp manufacturer for guidance. The data sheet for

your op amp will have a *voltage offset null circuit* recommended by the manufacturer. Experts have designed the null circuit to minimize offset errors at the
lowest cost to the user (see Appendices 1 and 2).

Some typical output-voltage null circuits are shown in Fig. 9-9. In Fig.
9-9(a), one variable resistor is connected between the $+V$ supply and a *trim*
terminal. For an expensive op amp, the manufacturer may furnish a metal film
resistor selected especially for that op amp. In Fig. 9-9(b), a 10-kΩ pot is connected between terminals called *offset null*. More complicated null circuits are

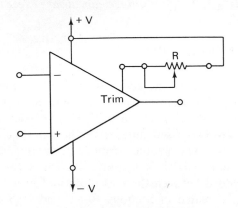

(a) Trim resistor 0 to 50 kΩ or 25 kΩ
fixed for discrete op amp

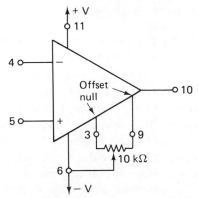

(b) 741 Offset voltage adjustment (DIP)

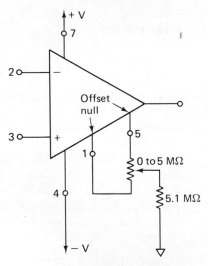

(c) 301 or 748 offset voltage
adjustment (TO 99 case)

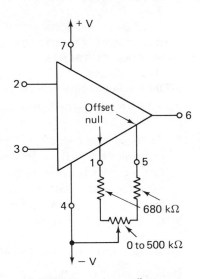

(d) 537 Offset voltage adjustment

Figure 9-9 Typical circuits to minimize error in output voltage due to input
offset voltage (see also Appendices 1 and 2).

shown in Fig. 9-9(c) and (d). Note that only the offset-voltage compensating resistors are shown by the manufacturer. They assume that a current-compensating resistor will be installed in series with the $(+)$ input.

9-7.3 Nulling Procedure for Output Voltage

1. Build the circuit. Include (a) the current-compensating resistor (see Section 9-4.3) and (b) the voltage offset null circuit (see Section 9-7.2).
2. Reduce all generator signals to 0. If their output cannot be set to 0, replace them with resistors equal to their internal resistance. This step is unnecessary if their internal resistance is negligible with respect to (more than about 1% of) any series resistor R_i connected to the generator.
3. Connect the load to the output terminal.
4. Turn on the power and wait a few minutes for things to settle down.
5. Connect a dc voltmeter or a CRO (dc coupled) across the load to measure V_o. (The voltage sensitivity should be capable of reading down to a few millivolts.)
6. Vary the offset voltage adjustment resistor until V_o reads 0 V. Note that output voltage errors due to both input offset voltage and input offset current are now minimized.
7. Install the signal sources and do *not* touch the offset-voltage adjustment resistor again.

9-8 DRIFT

It has been shown in this chapter that dc error components in V_0 can be minimized by installing a current-compensating resistor in series with the $(+)$ input and by trimming the offset-voltage adjustment resistor. It must also be emphasized that the zeroing procedure holds only at one temperature and at one time.

The offset current and offset voltage change with time because of aging of components. The offsets will also be changed by temperature changes in the op amp. In addition, if the supply voltage changes, bias currents, and consequently the offset current, change. By use of a well-regulated power supply, the output changes that depend on supply voltage can be eliminated. However, the offset changes with temperature can only be minimized by (1) holding the temperature surrounding the circuit constant, or (2) selecting op amps with offset current and offset voltage ratings that change very little with temperature changes.

The changes in offset current and offset voltage due to temperature are described by the term *drift*. Drift is specified for offset current in nA/°C (nanoamperes per degree Celsius). For offset voltage, drift is specified in μV/°C (microvolts per degree Celsius). Drift rates may differ at different temperatures and may even reverse; that is, at low temperatures V_{io} may drift by $+20$ μV/°C (increase), and at high temperatures V_{io} may change by -10 μV/°C (decrease). For this reason, manufacturers may specify either an average or maximum drift between two temperature limits. Even better is to have a plot of drift vs. temperature. An example is shown to calculate the effects of drift.

Example 9-6

A 301 op amp in the circuit of Fig. 9-10 has the following drift specifications. As temperature changes from 25°C to 75°C, I_{os} changes by a *maximum* of 0.3 nA/°C and V_{io} changes by a *maximum* of 30 μV/°C. Assume that V_o has been zeroed at 25°C; then the surrounding temperature is raised to 75°C. Find the maximum error in output voltage due to drift in (a) V_{io} and (b) I_{os}.

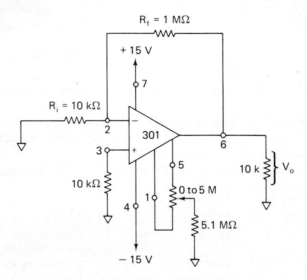

Figure 9-10   Circuit for Example 9-6.

Solution. (a) V_{io} will change by

$$\pm \frac{30\ \mu V}{°C} \times (75 - 25)°C = \pm 1.5\ mV$$

From Fig. 9-8(b), the change in V_o due to the change in V_{io} is

$$1.5\ mV\left(1 + \frac{R_f}{R_i}\right) = 1.5\ mV(101) \cong \pm 150\ mV$$

(b) I_{os} will change by

$$\pm \frac{0.3\ nA}{°C} \times 50°C = \pm 15\ nA$$

From Fig. 9-5(b), the change in V_o due to the change in I_{os} is $\pm 15\ nA \times R_f = \pm 15\ nA(1\ M\Omega) = \pm 15\ mV$.

The changes in V_o due to both V_{io} and I_{os} can either add or subtract from one another. Therefore, the worst possible change in V_o is either $+165\,\text{mV}$ or $-165\,\text{mV}$, from the 0 value at 25°C.

9-9   MEASUREMENT OF OFFSET VOLTAGE AND BIAS CURRENTS

The effects of offset voltage and bias currents have been discussed separately to simplify the problem of understanding how error voltage components appear in the dc output voltage of an op amp. However, their effects are *always* present simultaneously.

In order to measure V_{io}, I_{B+}, and I_{B-} of general-purpose op amps as inexpensively as possible, the following procedure is recommended.

1. As shown in Fig. 9-11(a), measure V_o with a digital voltmeter and calculate input offset voltage V_{io}:

$$V_{io} = \frac{V_o}{(R_f + R_i)/R_i} = \frac{V_o}{101} \tag{9-6}$$

Note that R_i and R_f are small. Therefore, by adding the 50-Ω

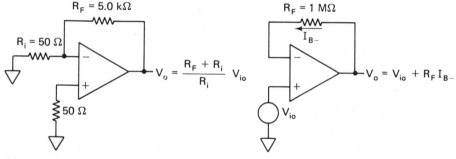

(a) Circuit to measure V_{io}; effect of I_{os} is minimized

(b) Circuit to measure I_{B-}

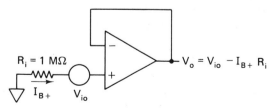

(c) Circuit to measure I_{B+}

Figure 9-11   Procedure to measure offset voltage, then bias currents for a general-purpose op amp.

current-compensation resistor, we force the output voltage error component due to I_{os} to be negligible.

2. To measure I_{B-}, set up the circuit of Fig. 9-11(b). Measure V_o. Using the value of V_{io} found in step 1, calculate I_{B-} from

$$I_{B-} = \frac{V_o - V_{io}}{R_f} \qquad (9\text{-}7)$$

3. To measure I_{B+}, measure V_o in Fig. 9-11(c) and calculate I_{B+} from

$$I_{B+} = -\left(\frac{V_o - V_{io}}{R_f}\right) \qquad (9\text{-}8)$$

Example 9-7

The circuit of Fig. 9-11 is used with the resistance values shown for a 741 op amp. Results are $V_o = +0.421$ V for Fig. 9-11(a), $V_o = 0.097$ V for Fig. 9-11(b), and $V_o = -0.082$ V for Fig. 9-11(c). Find (a) V_{io}; (b) I_{B-}; (c) I_{B+}.

Solution. (a) From Eq. (9-6),

$$V_{io} = \frac{0.421 \text{ V}}{101} = 4.1 \text{ mV}$$

(b) From Eq. (9-7),

$$I_{B-} = \frac{(97 - 4.1) \text{ mV}}{1 \text{ M}\Omega} = 93 \text{ nA}$$

(c) From Eq. (9-8),

$$I_{B+} = -\frac{(-82 - 4.1) \text{ mV}}{1 \text{ M}\Omega} = 86 \text{ nA}$$

Note that I_{os} is found to be $I_{B+} - I_{B-} = -7$ nA.

PROBLEMS

9-1. Which op amp characteristics normally have the most effect on (a) dc amplifier performance; (b) ac amplifier performance?

9-2. If $I_{B+} = 0.2 \ \mu$A and $I_{B-} = 0.1 \ \mu$A, find (a) the average bias current I_B; (b) the offset current I_{os}.

9-3. In Example 9-2, $V_o = -0.2$ V. Find I_{B-}.

9-4. In Example 9-3, $V_o = 0.2$ V. Find I_{B+}.

9-5. I_{B-} is 0.2 μA in Fig. 9-3(c). Find V_o.

9-6. In Fig. 9-5(a), $R_f = R_g = 100$ kΩ. $I_{B+} = 0.3 \ \mu$A and $I_{B-} = 0.2 \ \mu$A. Find V_o.

9-7. In Fig. 9-5(b), $R_f = R_i = 25$ kΩ and $R = 12.5$ kΩ. If $I_{os} = 0.1 \ \mu$A, find V_o.

9-8. In Fig. 9-5(b), $R_i = R_f = 25$ kΩ and $R = 12.5$ kΩ. If $I_{os} = -0.1 \ \mu$A, find V_o.

9-9. $V_o = 200$ mV in Fig. 9-7(b). Find V_{io}.

9-10. Resistors R_1, R_2, R_3, and R_f all equal 20 kΩ in Fig. 9-8(a). $E_1 = E_2 = E_3 = V_{io} = 2$ mV. Find (a) the actual value of V_o; (b) V_o assuming that $V_{io} = 0$.

9-11. What value of current-compensating resistor should be added in Problem 9-10?

9-12. What is the general procedure to null the output voltage of an op amp to 0 V?

9-13. In Fig. 9-10, V_{io} changes by ± 1 mV when the temperature changes by 50°C. What is the change in V_o due to the change in V_{io}?

9-14. I_{os} changes by ± 20 nA in Fig. 9-10 for a temperature change of 50°C. What is the resulting change in V_o?

9-15. $V_o = 101$ mV in the circuit of Fig. 9-11(a), $V_o = 201$ mV in Fig. 9-11(b), and $V_o = -99$ mV in Fig. 9-11(c). Find (a) V_{io}; (b) I_{B-}; (c) I_{B+}.

bandwidth, slew rate, noise, and frequency compensation

10

10-0 INTRODUCTION

When the op amp is used in a circuit that amplifies only ac signals, we must consider whether ac output voltages will be small signals (below about 1 V peak) or large signals (above 1 V peak). If only small ac output signals are present, the important op amp characteristics that limit performance are *noise* and *frequency response*. If large ac output signals are expected, then an op amp characteristic called *slew-rate limiting* determines whether distortion will be introduced by the op amp.

Bias currents and offset voltages affect dc performance and usually do not have to be considered with respect to ac performance. This is true because a coupling capacitor is usually in the circuit to pass ac signals and block dc currents and voltages. We begin with an introduction to the frequency response of an op amp.

10-1 FREQUENCY RESPONSE OF THE OP AMP

10-1.1 Internal Frequency Compensation

Many types of general-purpose op amps and specialized op amps are *internally compensated*; that is, the manufacturer has installed within such op amps a small capacitor, usually 30 pF. This *internal frequency compensation capacitor* prevents the op amp from oscillating at high frequencies. Oscillations are prevented by decreasing the op amp's gain as frequency increases. Otherwise, there would be sufficient gain and phase shift at some high frequency where enough output signal could be fed back to the input and cause oscillations (see Appendix 1).

From basic circuit theory it is known that the reactance of a capacitor goes down as frequency goes up: $X_C = 1/(2\pi f C)$. For example, if the frequency is increased by 10, the capacitor reactance decreases by 10. Thus, it is no accident that the voltage gain of an op amp goes down by 10 as the frequency of the input signal is increased by 10. A change in frequency of 10 is called a *decade*. Manufacturers show how the open-loop gain of the op amp is related to the frequency of the differential input signal by a curve called *open-loop voltage gain vs. frequency*. The curve may also be called *small-signal response*.

10-1.2 Frequency-Response Curve

A typical curve is shown in Fig. 10-1 for internally compensated op amps such as the 741 and 747. At low frequencies (below 0.1 Hz), the open-loop voltage gain is very high. A typical value is 200,000 (106 dB), and it is this value

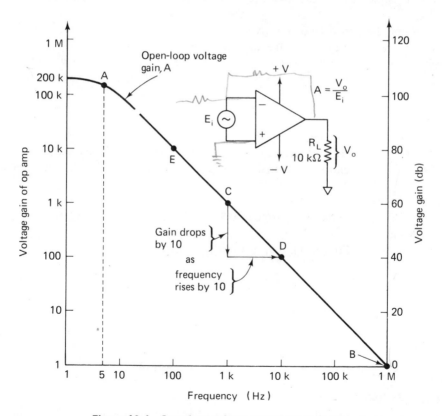

Figure 10-1 Open-loop voltage gain versus frequency.

that is specified on data sheets where a curve is not given. See "Large-Signal Voltage Gain" equals 200,000 in Appendix 1 and 160 V/mV in Appendix 2.

Point *A* in Fig. 10-1 locates the *break frequency* where the voltage gain is 0.707 times its value at very low frequencies. Therefore, the voltage gain at point *A* (where the frequency of E_i is 5 Hz) is about 140,000, or $0.707 \times 200,000$.

Points *C* and *D* show how gain drops by a factor of 10 as frequency rises by a factor of 10. Changing frequency or gain by a factor of 10 is expressed more efficiently by the term *per decade* ("decade" signifies 10). The right-hand vertical axis of Fig. 10-1 is a plot of voltage gain in decibels (dB). The voltage gain decreases by 20 dB for an increase in frequency of 1 decade. This explains why the frequency-response curve from *A* to *B* is described as *rolling off at 20 dB/ decade*. An alternative description is *6 dB/octave rolloff* ("octave" signifies a frequency change of 2). Therefore, each time the frequency doubles, the voltage gain decreases by 6 dB.

10-1.3 Unity-Gain Bandwidth

Point *B* in Fig. 10-1 defines the *small-signal unity-gain bandwidth* of the op amp. It is located at that frequency where the open-loop voltage gain is unity, or 1.

Some data sheets do not give a specification called unity-gain bandwidth or a curve like Fig. 10-1. Instead, they give a specification called *transient response rise time (unity gain)*. For a 741 op amp it is typically 0.25 μs and 0.8 μs at maximum. The bandwidth *B* is calculated from the rise-time specification by

$$B = \frac{0.35}{\text{rise time}} \tag{10-1}$$

where *B* is in hertz and rise time is in seconds. Rise time is defined in Section 10-1.4. (See Appendix 1, "Transient Response—Unity Gain" = 0.3 μs typical.)

Example 10-1

A 741 op amp has a rise time of 0.35 μs. Find the small-signal or unity-gain bandwidth.
Solution. From Eq. (10-1),

$$B = \frac{0.35}{0.35 \ \mu\text{s}} = 1 \text{ MHz}$$

Example 10-2

What is the open-loop voltage gain for the op amp of Example 10-1 at 1 MHz?
Solution. From the definition of *B*, the voltage gain is 1.

Example 10-3

What is the open-loop voltage gain at 100 kHz for the op amp in Examples 10-1 and 10-2?

Solution. By inspection of Fig. 10-1, if the frequency goes down by 10, the gain goes up by 10. Therefore, since the frequency goes down a decade (from 1 MHz to 100 kHz), the gain must go up by a decade from 1 at 1 MHz to 10 at 100 kHz.

Example 10-3 leads to the conclusion that if you divide the frequency of the signal, f, into the unity-gain bandwidth, B, the result is the op amp's gain at the signal frequency. Expressed mathematically,

$$\text{open-loop gain at } f = \frac{\text{bandwidth at unity gain}}{\text{input signal frequency,} f}. \qquad (10\text{-}2)$$

Example 10-4

What is the open-loop gain of an op amp that has a unity-gain bandwidth of 1.5 MHz for a signal of 1 kHz?

Solution. From Eq. (10-2), the open-loop gain at 1 kHz is

$$\frac{1.5 \text{ MHz}}{1 \text{ kHz}} = 1500$$

The data shown in Fig. 10-1 are useful for learning but will probably not apply to your op amp. For example, while 200,000 is a specified typical open-loop gain, the manufacturer guarantees only a minimum gain of 20,000 for general-purpose op amps. Still, 20,000 may be enough to do the job. Section 10-2 deals with this question.

10-1.4 Rise Time

Assume that the input voltage E_i of a unity-gain amplifier is changed very rapidly by a square wave or pulse signal. Ideally, E_i should be changed from 0 V + 20 mV in 0 time; practically, a few nano-seconds are required to make this change (see Appendix 1, "Transient Response Curve"). At unity gain, the output should change from 0 to +20 mV in the same few nanoseconds. However, it takes time for the signal to propagate through all the transistors in the op amp. It also takes time for the output voltage to rise to its final value. *Rise time* is defined as the time required for the output voltage to rise from 10% of its final value to 90% of its final value. From Section 10-1.3, the rise time of a 741 is 0.35 μs. Therefore, it would take 0.35 μs for the output voltage to change from 2 mV to 18 mV.

10-2 AMPLIFIER GAIN AND FREQUENCY RESPONSE

10-2.1 Effect of Open-Loop Gain on Closed-Loop Gain of an Amplifier, DC Operation

It is necessary to learn how open-loop gain A_{OL} affects the actual closed-loop gain of an amplifier with dc signal voltages (zero frequency). First, we must define *ideal* closed-loop gain of an amplifier as that gain which should be determined only by external resistors. However, the *actual* closed loop of an amplifier is determined by *both* the external resistors and open-loop gain of an op amp.

The actual closed-loop gain of a *noninverting* amplifier is

$$\text{actual } A_{CL} = \frac{\text{ideal } A_{CL}}{1 + \text{ideal } A_{CL}/A_{OL}} \qquad \text{(10-3a)}$$

where ideal $A_{CL} = (R_f + R_i)/R_i$.

For an *inverting amplifier* the actual gain is

$$\text{actual } A_{CL} = \frac{\text{ideal } A_{CL}}{1 + (1 - \text{ideal } A_{CL})/A_{OL}} \qquad \text{(10-3b)}$$

where ideal $A_{CL} = -R_f/R_i$.

Example 10-5

Find the actual gain for a dc noninverting amplifier if ideal $A_{CL} = 100$ and A_{OL} is (a) 10,000; (b) 1000; (c) 100; (d) 10; (e) 1. Repeat for a dc inverting amplifier with an ideal gain of -100.

Solution. (a) From Eq. (10-3a), noninverting

$$\text{actual } A_{CL} = \frac{100}{1 + 100/10,000} = 99.0099$$

From Eq. (10-3b), the inverting

$$\text{actual } A_{CL} = \frac{-100}{1 + (1 + 100)/10,000} = -99.0000$$

Tabulate the results:

				A_{OL}		
	1	10	100	10^3	10^4	10^5
Actual A_{CL}, noninverting	0.99	9.0	50	90.9	99.0	99.9
Actual A_{CL}, inverting	−0.98	−9.0	−49.7	−90.9	−99	−99.9

The results of Example 10-5 are shown by the plot of A_{CL} versus A_{OL} in Fig. 10-2. There are two important lessons to be learned from Example 10-5 and Fig. 10-2. First the actual gains of both noninverting and inverting amplifiers are of approximately the same magnitudes for the same value of open-loop gain. Second, we would like the actual closed-loop gain to be equal to the ideal closed-loop gain. An examination of Eqs. (10-3a) and (10-3b) shows that this will be true *if* the open-loop gain of the op amp A_{OL} is large with respect to the

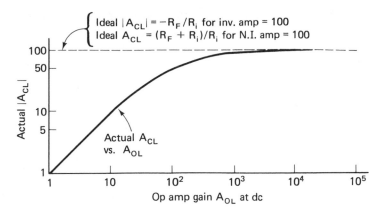

Figure 10-2 The actual closed-loop dc gain of an inverting or noninverting amplifier depends on both the ideal gain that is set by resistor ratios, and the open-loop gain of the op amp at dc.

ideal closed-loop gain of the op amp. Practically, we would like A_{OL} to be 100 or more times the ideal A_{CL}, so that the external precision resistors and *not* the op amp's A_{OL} determine the actual gain.

We already learned in Section 10-1.2 that A_{OL} depends on frequency. Since A_{OL} of the op amp also determines A_{CL} of an amplifier, then A_{CL} of the amplifier will also depend on frequency.

10-2.2 Gain–Bandwidth Product

One figure of merit for an op amp is its gain–bandwidth product (GB). GB is found by multiplying the open-loop gain of the op amp at low frequencies (below 0.1 Hz) by its bandwidth (see Section 10-3 for a definition of bandwidth). For example, the GB of a 741 is found by reference to Fig. 10-1. $A_{OL} = 200,000$ and bandwidth is 5 Hz; thus GB = 200,000 × 5 Hz = 1 MHz. Note also that the op amp has an $A_{OL} = 1$ with a (unity gain) bandwidth of 1 MHz, or gain bandwidth = 1 × 1 MHz = 1 MHz. *Thus the unity gain–bandwidth of the op amp* (Section 10-1.3) *equals its gain–bandwidth product.*

When the op amp is used to make an amplifier, *the gain–bandwidth product of the amplifier is also equal to the gain–bandwidth product of the op amp.* Expressed mathematically,

$$A_{CL} \times \text{bandwidth of amplifier} = A_{OL} \times \text{bandwidth of op amp} \quad \text{(10-3c)}$$

This relationship is applied to amplifier circuits in the next section.

10-2.3 Small-Signal Bandwidth

The useful frequency range of any amplifier (closed- or open-loop) is defined by a high-frequency limit f_H and a low-frequency limit f_L. At f_L and f_H, the voltage gain is down to 0.707 times its maximum value in the middle of the useful frequency range. In terms of decibels, the voltage gain is down 3 dB at both f_L and f_H. These statements are summarized on the general frequency-response curve in Fig. 10-3 and in Appendices 1 and 2.

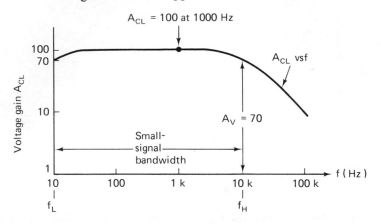

Figure 10-3   Small-signal bandwidth.

Small-signal bandwidth is the difference between f_H and f_L. Often, f_L is very small with respect to f_H, or f_L is 0 for a dc amplifier. Therefore, the small-signal bandwidth approximately equals the high-frequency limit f_H. From point *A* of Fig. 10-1, we see that the small-signal bandwidth of an op amp is 5 Hz.

The small-signal bandwidth for closed-loop amplifiers is determined by both the unity-gain bandwidth (Section 10-1.3) *B* and the closed-loop gain A_{CL}. The relationship is simply

$$\text{closed-loop small-signal bandwidth} = \frac{B}{A_{CL}} \qquad (10\text{-}4)$$

Equation (10-4) can be shown graphically as in Fig. 10-4. Calculate the ideal A_{CL} from $R_f/R_i = 1000$. Extend the horizontal lines of Ideal A_{CL} to intersect the open-loop curve at point f_H. Read the small-signal bandwidth directly below f_H as 1 kHz. Observe that the closed-loop gain at f_H is $0.707 \times$ ideal gain $= 0.707 \times 1000 = 700$. For frequencies above f_H, the gain is determined not by R_f and R_i but by the op amp.

Example 10-6

The unity-gain bandwidth of a 741 is 1 MHz. When it is used in an amplifier designed for a closed-loop gain of 100, find (a) the small-signal bandwidth; (b) A_{CL} at f_H; (c) the same quantities for an amplifier with a closed-loop gain of 10.

Solution. (a) By Eq. (10-4),

$$\text{bandwidth} = \frac{10^6 \text{ Hz}}{100} = 10 \text{ kHz}$$

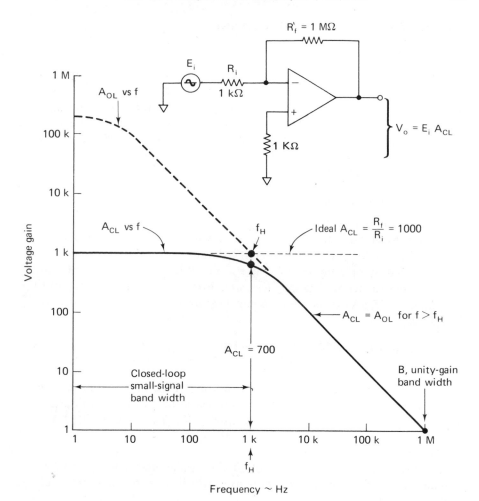

Figure 10-4 Small-signal bandwidth and closed-loop gain.

(b) A_{CL} at $f_H = 100 \times 0.707 = 70$.
(c) By Eq. (10-4),

$$\text{bandwidth} = \frac{10^6 \text{ Hz}}{10} = 100 \text{ kHz}$$

A_{CL} at $f_H = 10 \times 0.707 \cong 7$.

Example 10-6 and Eq. (10-4) show that there is a direct trade-off between small-signal gain and bandwidth. If you increase the gain by 10, the bandwidth decreases by 10. Notice that the product of closed-loop gain and small-signal bandwidth always equals the unity-gain bandwidth B. For this reason, the unity-gain bandwidth is also called *gain–bandwidth product* and is a figure of merit for the op amp.

10-3 SLEW RATE AND OUTPUT VOLTAGE

10-3.1 Definition of Slew Rate

The slew rate of an op amp tells how fast its output voltage can change. For a general-purpose op amp such as the 741, the maximum slew rate is 0.5 V/μs. This means that the output voltage can change a maximum of $\frac{1}{2}$ V in 1 μs. Slew rate depends on many factors: the amplifier gain, compensating capacitors, and even whether the output voltage is going positive or negative. The worst case, or slowest slew rate, occurs at unity gain. Therefore, slew rate is usually specified at unity gain (see Appendices 1 and 2).

10-3.2 Cause of Slew-Rate Limiting

Either within or outside the op amp there is at least one capacitor required to prevent oscillation (see Section 10-1.1). Connected to this capacitor is a portion of the op amp's internal circuitry that can furnish a maximum current that is limited by op amp design. The ratio of this maximum current I to the compensating capacitor C is the *slew rate*. For example, a 741 can furnish a maximum of 15 μA to its 30-pF compensating capacitor (see Appendix 1). Therefore,

$$\text{slew rate} = \frac{\text{output voltage change}}{\text{time}} = \frac{I}{C} = \frac{15 \ \mu\text{A}}{30 \ \text{pF}} = 0.5 \ \frac{\text{V}}{\mu\text{s}} \quad (10\text{-}5)$$

From Eq. (10-5), a faster slew rate requires the op amp to have either a higher maximum current or a smaller compensating capacitor. For example, the AD518 has a slew rate of 80 V/μs with $I = 400 \ \mu$A and $C = 50$ pF.

Example 10-7

An instantaneous input change of 10 V is applied to a unity-gain inverting amplifier. If the op amp is a 741, how long will it take for the output voltage to change by 10 V?
Solution. By Eq. (10-5),

$$\text{slew rate} = \frac{\text{output voltage change}}{\text{time}}$$

$$\frac{0.5 \ \text{V}}{\mu\text{s}} = \frac{10 \ \text{V}}{\text{time}}, \qquad \text{time} = \frac{10 \ \text{V} \times \mu\text{s}}{0.5 \ \text{V}} = 20 \ \mu\text{s}$$

10-3.3 Slew-Rate Limiting of Sine Waves

In the voltage follower of Fig. 10-5, E_i is a sine wave with peak amplitude E_p. The maximum rate of change of E_i depends on both its frequency f and the peak amplitude as given by $2\pi f E_p$. If this rate of change is larger than the op amp's slew rate, the output V_o will be distorted. That is, output V_o tries to follow E_i but cannot do so because of slew-rate limiting. The result is distortion, as shown by the triangular shape of V_o in Fig. 10-5. The maximum frequency f_{max}

at which we can obtain an undistorted output voltage with a peak value of V_{op} is determined by the slew rate in accordance with

$$f_{max} = \frac{\text{slew rate}}{6.28 \times V_{op}} \tag{10-6}$$

where f_{max} is the maximum frequency in Hz, V_{op} is the maximum undistorted output voltage in volts, and the slew rate is in volts per microsecond.

Example 10-8

The slew rate for a 741 is 0.5 V/μs. At what maximum frequency can you get an undistorted output voltage of (a) 10 V peak; (b) 1 V peak?

Solution. (a) From Eq. (10-6),

$$f_{max} = \frac{1}{6.28 \times 10 \text{ V}} \times \frac{0.5 \text{ V}}{\mu s} = 8 \text{ kHz}$$

(b) From Eq. (10-6),

$$f_{max} = 80 \text{ kHz}$$

Example 10-9

The peak output voltage that can be gotten out of a 741 is 13 V with $\pm$15-V supply. This output voltage is described as *full power output*. What is the maximum frequency for full power output of a 741? Note this *full power output frequency* specification is often supplied by the manufacturer (see Appendix 1, "Output Voltage Swing as a Function of Frequency").

Solution. From Eq. (10-6),

$$f_{max \text{ full power}} = \frac{1}{6.28 \times 13 \text{ V}} \times \frac{0.5 \text{ V}}{\mu s} = 6 \text{ kHz}$$

Examples 10-8 and 10-9 show that the slew rate limits the upper frequency of large-amplitude output voltages. As the peak output voltage required from the op amp is reduced, the upper-frequency limitation imposed by the slew rate increases.

Recall that the upper-frequency limitation imposed by small-signal response increases as the closed-loop gain decreases. For each amplifier application, the upper-frequency limit imposed by slew-rate limiting (Section 10-3.3) and small-signal bandwidth (Section 10-2.3) must be calculated. The *smaller* value determines the actual upper-frequency limit. In general, the slew rate is a large-signal frequency limitation and small-signal frequency response is a small-signal frequency limitation.

10-3.4 Slew Rate Made Easy

Figure 10-6 simplifies the problem of finding f_{max} at any peak output voltage for slew rates between 0.5 and 5 V/μs. For example, to do part (a) of Example

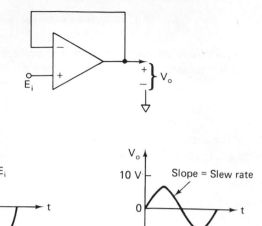

Figure 10-5 Example of slew-rate limiting of output voltage V_o.

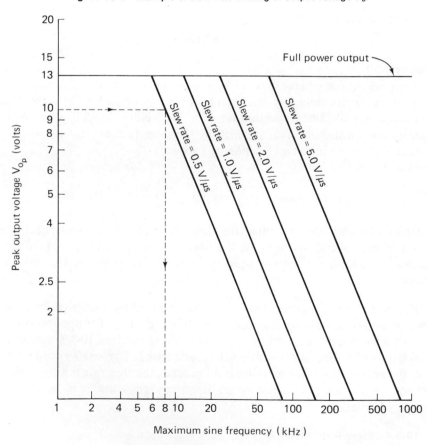

Figure 10-6 Slew rate made easy. Any point on a slew-rate line shows the maximum sinusoidal frequency allowed for the corresponding peak output voltage.

10-8, locate where the horizontal line of $V_{op} = 10$ V intersects the slew-rate line 0.5 V/μs. Below the intersection, read $f_{max} = 8$ kHz.

10-4 NOISE IN THE OUTPUT VOLTAGE

10-4.1 Introduction

Undesired electrical signals present in the output voltage are classified as *noise*. Drift (see Chapter 9) and offsets can be considered as very low frequency noise. If you view the output voltage of an op amp amplifier with a sensitive CRO (1 mV/cm), you will see a random display of noise voltages called *hash*. The frequencies of these noise voltages range from 0.01 Hz to megahertz.

Noise is generated in any material that is above absolute zero ($-273°$C). Noise is also generated by all electrical devices and their controls. For example, in an automobile, the spark plugs, voltage regulator, fan motor, air conditioner, and generator all generate noise. Even when headlights are switched on (or off), there is a sudden change in current that generates noise. This type of noise is external to the op amp. Effects of external noise can be minimized by proper construction techniques and circuit selection (see Section 10-4.3 to 10-4.5).

10-4.2 Noise in Op Amp Circuits

Even if there were no external noise, there would still be noise in the output voltage caused by the op amp. This internal op amp noise is modeled most simply by a noise voltage source E_n. As shown in Fig. 10-7, E_n is placed in series with the (+) input. On data sheets, noise voltage is specified in microvolts (rms)

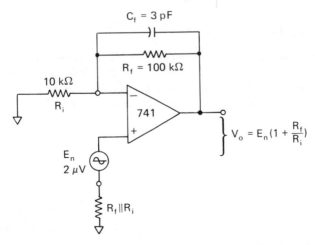

Figure 10-7 Op amp noise is modeled by a noise voltage in series with the (+) input.

for different values of source resistance over a particular frequency range. For example, the 741 op amp has 2 μV of *total* noise over a frequency of 10 Hz to 10 kHz. This noise voltage is valid for source resistors (R_i) between 100 Ω and 20 kΩ. The noise voltage goes up directly with R_i, once R_i exceeds 20 kΩ. Thus R_i should be kept below 20 kΩ to minimize noise in the output (see Appendix 1).

10-4.3   Noise Gain

Noise voltage is amplified just as offset voltage is. That is, *noise voltage gain* is the same as the gain of a non-inverting amplifier:

$$\text{noise gain} = 1 + \frac{R_f}{R_i}$$

What can you do about minimizing output voltage errors due to noise? First, avoid, if possible, large value of R_i and R_f. Install a small capacitor (3-pF) across R_f to shunt it at high noise frequencies. Then the higher noise frequencies will not be amplified so much. Next, do not shunt R_i with a capacitor; otherwise, the $R_i C$ combination will have a smaller impedance at higher noise frequencies than R_i alone, and gain will increase with frequency and aggravate the situation. Finally, try to keep R_i at about 10 kΩ or below.

Noise currents, like bias currents, are also present at each op amp input terminal. If a bias-current compensation resistor is installed (see Chapter 9), the effect of noise currents on output voltage will be reduced.

As with offset current, the effects of noise currents also depend on the feed-back resistor. So if possible, reduce the size of R_f to minimize the effects of noise currents.

10-4.4   Noise in the Inverting Adder

In the inverting adder (see Section 3-2), each signal input voltage has a gain of 1. However, the noise gain will be 1 plus the number of inputs; for example, a four-input adder would have a noise gain of 5. Thus noise voltage has five times as much gain as each input signal. Therefore, low amplitude signals should be preamplified before connecting them to an adder.

10-4.5   Summary

To reduce the effects of op amp noise:

1. *Never* connect a capacitor across the input resistor or from ($-$) input to ground. There will always be a few pF of stray capacitance from ($-$) input to ground due to wiring, so

2. *Always* connect a small capacitor (3 pF) across the feedback resistor. This reduces the noise gain at high frequencies.
3. If possible, avoid large resistor values.

10-5 EXTERNAL FREQUENCY COMPENSATION

10-5.1 Need for External Frequency Compensation

Op amps with internal frequency compensation (see Section 10-1.1 and Appendix 1) are very stable with respect to signal frequencies. They do not burst into spontaneous oscillation or wait to oscillate occasionally when a signal is applied. However, the trade-offs for frequency stability are limited small-signal bandwidth, slow slew rate, and reduced-power bandwidth. Internally compensated op amps are useful at audio frequencies but not at higher frequencies.

A 741 that has a 1-MHz gain–bandwidth product will give a useful gain of 1000 only up to frequencies of about 1 kHz. To obtain more gain from the op amp at higher frequencies, the internal frequency-compensating capacitor of the op amp must removed. If this is done, the resulting op amp structure has a higher slew rate and greater power bandwidth. But these inprovements would be cancelled out because the op amp probably would oscillate continually. As usual, there is a trade-off: frequency stability for a larger bandwidth and slew rate.

In order to be able to make these trade-offs, manufacturers of op amps bring out from one to three *frequency-compensating terminals*. Such terminals allow the user to choose the best allowable combination of stability and bandwidth. This choice is made by connecting external capacitors and resistors to the compensating terminals. Accordingly, this versatile type of op amp is classified as *externally frequency-compensated* (see Appendix 2).

10-5.2 Single-Capacitor Compensation

The frequency response of the 101 general-purpose op amp can be tailored by connecting a single capacitor, C_1, to pins 1 and 8. As shown in Fig. 10-8 and Appendix 2, by making $C_1 = 3$ pF, the 101 has an open-loop frequency-response curve with a small-signal bandwidth of 10 MHz. Increasing C_1 by a factor of 10 (to 30 pF) reduces the small-signal bandwidth by a factor of 10 (to 1 MHz). Therefore, the 101 can be externally compensated to have the same small-signal bandwidth as the 741.

When the 101 is used in an amplifier circuit, the amplifier's useful frequency range now depends on the compensating capacitor. For example, with R_f/R_i set for an amplifier gain of 100, its small-signal bandwidth would be 10 kHz for $C_1 = 30$ pF. By reducing C_1 to 3 pF, the small-signal bandwidth is increased to 100 kHz. The full-power bandwidth is also increased, from about 6 kHz to 60 kHz.

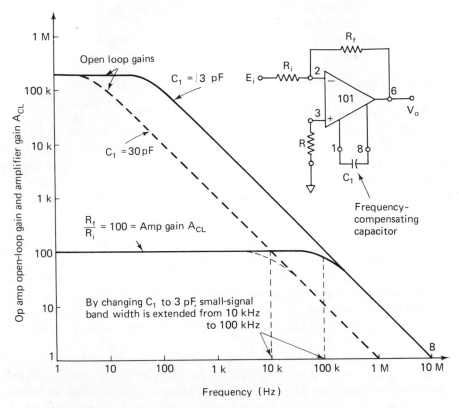

Figure 10-8 Extending frequency response with an external compensating capacitor.

10-5.3 Feed-Forward Frequency Compensation

There are many other types of frequency compensation. Among the more popular are *two-capacitor, or two-pole, compensation* and *feed-forward compensation*. Manufacturers' data sheets give precise instructions on the type best suited for your application.

Feed-forward compensation for the 101 is illustrated in Fig. 10-9. Feed-forward capacitor C is wired from the $(-)$ input to compensating terminal 1. A small capacitor C_f is needed across R_f to ensure frequency stability. The slew rate is increased to 10 V/μs and the full-power bandwidth to over 200 kHz. Of course, the added high-frequency gain will also amplify high-frequency noise.

We should conclude that frequency-compensation techniques must be applied only to the extent required for the circuit. Do not use any more high-frequency gain than is absolutely necessary; otherwise, there will be a needless amount of high-frequency noise in the output.

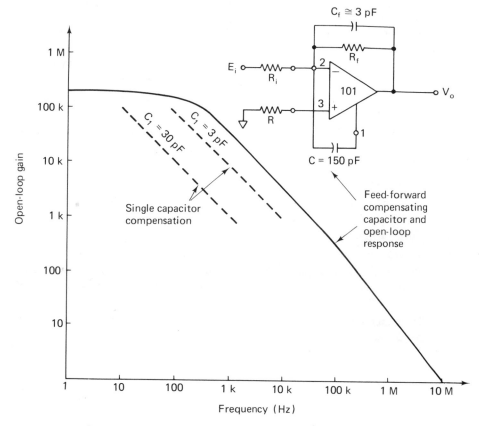

Figure 10-9 Extending bandwidth with feed-forward frequency compensation.

PROBLEMS

10-1. What is the typical open-loop gain of a 741 op amp at very low frequencies?

10-2. The dc open-loop gain of an op amp is 100,000. Find the open-loop gain at its break frequency.

10-3. The transient response rise time (unity gain) of an op amp is 0.07 μs. Find the small-signal bandwidth.

10-4. An op amp has a small-signal unity-gain bandwidth of 2 MHZ. Find its open-loop gain at 200 kHZ.

10-5. What is the difference between the open-loop and closed-loop gain of an op amp?

10-6. What is the open-loop gain for the op amp of Problem 10-4 at 2 MHZ?

10-7. What is the definition of "rise time"?

10-8. An op amp has a dc open-loop gain of 100,000. It is used in an inverting amplifier circuit with an ideal closed-loop gain of 1000. Find the actual dc closed-loop gain.

10-9. The op amp of Problem 10-8 is used in a noninverting amplifier with an ideal gain of 1000. Find the amplifier's actual dc closed-loop gain.

10-10. What is the small-signal bandwidth of the op amp whose frequency response is given in Fig. 10-1?

10-11. The unity-gain bandwidth of an op amp is 10 MHz. It is used to make an amplifier with an ideal closed-loop gain of 100. Find the amplifier's (a) small-signal bandwidth; (b) A_{CL} at f_H.

10-12. How fast can the output of an op amp change by 10 V if its slew rate is 1 V/μs?

10-13. Find the maximum frequency for a sine-wave output voltage of 10 V peak with an op amp whose slew rate is 1 V/μs.

10-14. Find the noise gain for an inverting amplifier with a gain of $R_F/R_i = -10$.

10-15. What is the noise for a five-input inverting adder?

10-16. The op amp in Example 10-9 is changed to one with a slew rate of 5 V/μs. Find its maximum full-power output frequency. Assume that $V_{o\ max} = 10$ V peak.

10-17. Does increasing the compensating capacitor increase or decrease unity-gain bandwidth?

modulating, demodulating, and frequency changing with the multiplier

ᑊᵾ

11

11-0 INTRODUCTION

Analog multipliers are complex arrangements of op amps and other circuit elements now available in either integrated circuit or functional module form. Multipliers are easy to use; some of their applications are (1) measurement of power, (2) frequency doubling and shifting, (3) detecting phase-angle difference between two signals of equal frequency, (4) multiplying two signals, (5) dividing one signal by another, (6) taking the square root of a signal, and (7) squaring a signal. Another use for multipliers is to demonstrate the principles of amplitude modulation and demodulation. The schematic of a typical multiplier is shown in Fig. 11-1(a). There are two input terminals, x and y, which are used for connecting the two voltages to be multiplied. Typical input resistance of each input terminal is 10 kΩ or greater. One output terminal furnishes about the same current as an op amp to a grounded load (5 to 10 mA). The output voltage equals the product of the input voltages reduced by a scale factor. The *scale factor* is explained in Section 11-1.

11-1 MULTIPLYING DC VOLTAGES

11-1.1 Multiplier Scale Factor

The schematic of a multiplier shown in Fig. 11-1(a) may have a $\times$ to symbolize multiplication. Another type of schematic shows the inputs and the output voltage equation, as in Fig. 11-1(b). In general terms, the output voltage

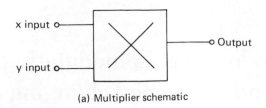

(a) Multiplier schematic

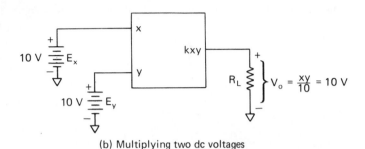

(b) Multiplying two dc voltages

Figure 11-1 Introduction to the multiplier.

V_o is the product of input voltages x and y and is expressed by

$$V_o = kxy \qquad (11\text{-}1a)$$

The constant k is called a *scale factor* and is usually equal to $\frac{1}{10}$. This is because multipliers are designed for the same type of power supplies used for op amps, namely ± 15 V. For best results, the voltages applied to either x or y inputs should not exceed $+10$ V or -10 V with respect to ground. This ± 10-V limit also holds for the output, so the scale factor is usually the reciprocal of the voltage limit, or $\frac{1}{10}$. If both input voltages are at their positive limits of $+10$ V, the output will be at its positive limit of 10 V. Thus Eq. (11-1a) is expressed for most multipliers by

$$V_o = \frac{xy}{10} = \frac{E_x E_y}{10} \qquad (11\text{-}1b)$$

11-1.2 Multiplier Quadrants

Multipliers are classified by quadrants; for example, there are one-quadrant, two-quadrant, and four-quadrant multipliers. The classification is explained in two ways in Fig. 11-2. In Fig. 11-2(a), the input voltages can have four possible polarity combinations. If both x and y are positive, operation is in quadrant 1, since x is the horizontal and y the vertical axis. If x is positive and y is negative, quadrant 4 operation results, and so forth.

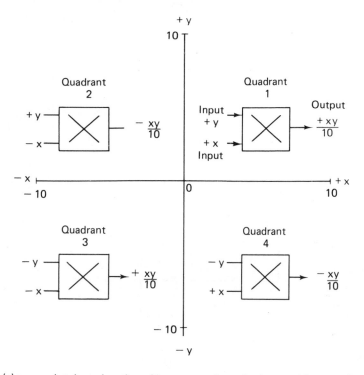

(a) y vs x plot shows location of input operating point in one of four quadrants

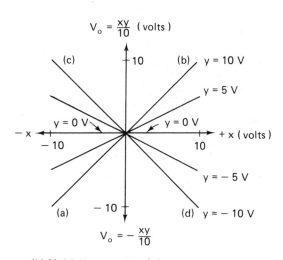

(b) Multiplier output xy/10 versus input x

Figure 11-2 Multiplying two dc voltages, *x* and *y*.

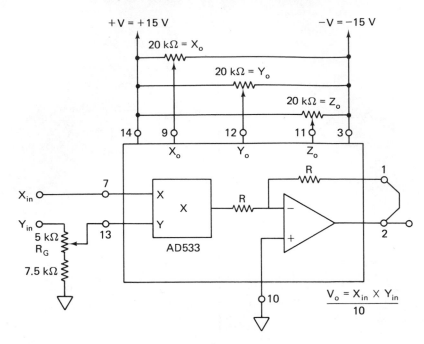

Trim Procedure

Step	Adjust pot	X_{in} at	Y_{in} at	For V_o =
1	Z_o	0 V	0 V	0 V dc
2	X_o	0 V	20 V, p-p, 50 Hz	Min. ac
3	Y_o	20 V, p-p, 50 Hz	0 V	Min. ac
4		Repeat steps 1 through 3, as required		
5	R_G	+10 V dc	20 V, p-p, 50 Hz	Y_{in}

Figure 11-3 Calibration procedure and circuit to trim an integrated-circuit multiplier.

Example 11-1

Find V_o for the following combination of inputs: (a) $x = 10$ V, $y = 10$ V; (b) $x = -10$ V, $y = 10$ V; (c) $x = 10$ V, $y = -10$ V; (d) $x = -10$ V, $y = -10$ V.

Solution. From Eq. (11-1b),

(a) $V_o = \dfrac{(10)(10)}{10} = 10$ V, quadrant 1

(b) $V_o = \dfrac{(-10)(10)}{10} = -10$ V, quadrant 2

(c) $V_o = \dfrac{(10)(-10)}{10} = -10$ V, quadrant 4

(d) $V_o = \dfrac{(-10)(-10)}{10} = 10$ V, quadrant 3

In Fig. 11-2(b), V_o is plotted on the vertical axis and x on the horizontal axis. If we apply 10 V to the y input and vary x from -10 V to $+10$ V, we plot the line ab labeled $y = 10$ V. If y is changed to -10 V, the line cd labeled $y = -10$ V results. These lines can be seen on a cathode-ray oscilloscope (CRO) by connecting V_o of the multiplier to the y input of the CRO and x of the multiplier to the $+x$ input of the CRO.

For accuracy, V_o should be 0 V when either multiplier input is 0 V. If this is not the case, a zero trim adjustment should be made, as described in Section 11-1.3.

11-1.3 Multiplier Calibration

Ordinary low-cost integrated-circuit multipliers such as the AD533, 4200, and XR2208 require manual calibration with external circuitry. Precision multipliers such as the AD534 require little or no manual calibration. Any internal unbalances are trimmed out precisely by the manufacturer, using computer-controlled lasers. Their cost is higher by a factor of 2 or 3.

The trim procedure for a popular low-cast IC multiplier is given in Fig. 11-3 together with the supporting circuitry. Three pots Z_0, X_0, and Y_0 are adjusted in sequence to give (1) 0 V out when both x and y inputs are 0 V, (2) 0 V out when the x input equals 0 V, and (3) 0 V out when the y input equals 0 V. Finally, the scale factor is adjusted to 0.1 by pot R_G (in step 5) to ensure that the output is at 10 V when both inputs are at 10 V.

11-2 SQUARING A NUMBER OR DC VOLTAGE

Any positive or negative number can be squared by a multiplier, providing that the number can be represented by a voltage between 0 and 10 V. Simply connect the voltage E_i to *both* inputs as shown in Fig. 11-4. This type of connection is known as a *squaring circuit*.

Example 11-2

Find V_o in Fig. 11-4 if (a) $E_i = +10$ V; (b) $E_i = -10$ V.
Solution. From Fig. 11-4,

(a) $V_o = \dfrac{10^2}{10} = 10$ V

(b) $V_o = \dfrac{(-10)(-10)}{10} = 10$ V

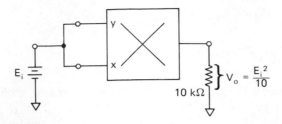

Figure 11-4 Squaring a dc voltage.

Example 11-2 shows that the output of the multiplier follows the rules of algebra; that is, when either a positive or negative number is squared, the result is a positive number.

11-3 FREQUENCY DOUBLING

11-3.1 Principle of the Frequency Doubler

An ideal frequency doubler would give an output voltage whose frequency is twice the frequency of the input voltage. The doubler circuit should not incorporate a tuned circuit, since the tuned circuit can be tuned only to one frequency. A true doubler should double any frequency. The multiplier is very nearly an ideal doubler if only one frequency is applied to both inputs. The output voltage for a doubler circuit is given by the trigonometric identity

$$(\sin 2\pi ft)^2 = \frac{1}{2} - \frac{\cos 2\pi(2f)t}{2} \qquad (11\text{-}2)$$

Equation (11-2) predicts that squaring a sine wave with a frequency of (for example) $f = 10$ kHz gives a negative cosine wave with a frequency of $2f$ or 20 kHz plus a dc term of $\frac{1}{2}$. Note that *any* input frequency f will be doubled when passed through a squaring circuit.

11-3.2 Squaring a Sinusoidal Voltage

In Fig. 11-5(a), sine-wave voltage E_i is applied to both multiplier inputs. E_i has a peak value of 5 V and a frequency of 10 kHz. The output voltage V_o is predicted by the calculations shown in Example 11-3.

Example 11-3

Calculate V_o in the squaring circuit or frequency doubler of Fig. 11-5.
Solution. Input $E_x = E_y = E_i$ and is expressed is volts by

$$E_i = E_x = E_y = 5 \sin 2\pi 10{,}000t$$

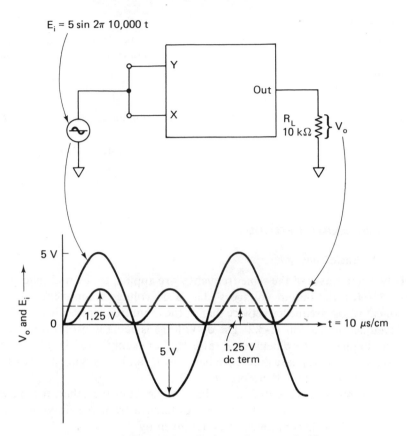

$E_i = 5 \sin 2\pi\, 10{,}000\, t$

Figure 11-5 Squaring circuit as a frequency doubler.

Substituting into Eq. (11-1b) yields

$$V_o = \frac{E_i^2}{10} = \frac{5^2}{10}(\sin 2\pi 10{,}000t)^2 \tag{11-3}$$

Applying Eq. (11-2), we obtain

$$V_o = 2.5\left[\frac{1}{2} - \frac{\cos 2\pi 20{,}000t}{2}\right]\text{V}$$

$$= \underbrace{1.25}\qquad\qquad - \underbrace{1.25 \cos 2\pi 20{,}000t}$$

= dc term of 1.25 V — frequency doubled to 20,000 Hz, 1.25 V peak

Both E_i and V_o are shown in Fig. 11-5. If you want to remove the dc voltage, simply install a 1-μF coupling capacitor between R_L and the output terminal. If you want to measure the dc voltage, simply connect a dc voltmeter to V_o.

Conclusion

V_o has two voltage components: (1) a dc voltage equal to $\frac{1}{20}\,(E_{i_p})^2$, and (2) an ac sinusoidal wave whose peak value is $\frac{1}{20}(E_{i_p})^2$ and whose frequency is double that of E_i.

Example 11-4

What are the dc and ac output voltage components of Fig. 11-5 if (a) $E_i = 10$ V peak at 1 kHz; (b) $E_i = 2$ V peak at 2.5 kHz?

Solution. (a) dc value $= (10)^2/20 = 5$ V; peak ac value $= (10)^2/20 = 5$ V at 2 kHz.

(b) dc value $= (2)^2/20 = 0.2$ V; peak ac value $= (2)^2/20 = 0.2$ V at 5 kHz.

11-4 PHASE-ANGLE DETECTION

11-4.1 Basic Theory

If two sine waves of the *same* frequency are applied to the multiplier inputs in Fig. 11-6(a), the output voltage V_o has a dc voltage component and an ac component whose frequency is twice that of the input frequency. This conclusion was developed in Section 11-3.2. The dc voltage is actually proportional to the difference in phase angle θ between E_x and E_y. For example, in Fig. 11-5, $\theta = 0°$, because there was no phase difference between E_x and E_y. Figure 11-6(b) shows a phase difference of $90°$; therefore, $\theta = 90°$.

If one input sine wave differs in phase angle from the other, it is possible to calculate or measure the phase-angle difference from the dc voltage component in V_o. This dc component $V_{o\,dc}$ is given by

$$V_{o\,dc} = \frac{E_{x_p}E_{y_p}}{20}(\cos\theta) \qquad (11\text{-}4a)$$

where E_{x_p} and E_{y_p} are peak amplitudes of E_x and E_y. For example, if $E_{x_p} = 10$ V, $E_{y_p} = 5$ V, *and they are in phase*, then $V_{o\,dc}$ would indicate 2.5 V on a dc voltmeter. This voltmeter point would be marked as a phase angle of $0°$ ($\cos 0° = 1$). If $\theta = 45°$ ($\cos 45° = 0.707$), the dc meter would read 0.707×2.5 V $\cong 1.75$ V. Our dc voltmeter can be calibrated as a *phase angle meter* $0°$ at 2.5 V, $45°$ at 1.75 V, and $90°$ at 0 V.

Equation (11-4a) may also be expressed by

$$\cos\theta = \frac{20V_{o\,dc}}{E_{x_p}E_{y_p}} \qquad (11\text{-}4b)$$

If we could arrange for the product $E_{x_p}\,E_{y_p}$ to equal 20, we could use a 0- to 1-V dc voltmeter to read $\cos\theta$ directly from the meter face and calibrate the

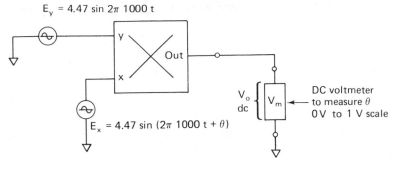

(a) Phase angle measurement

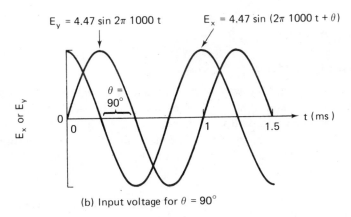

(b) Input voltage for $\theta = 90°$

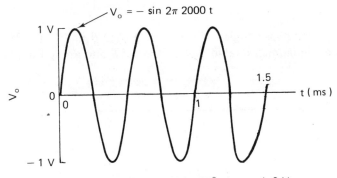

(c) Output voltage for $\theta = 90°$, dc term is 0 V

Figure 11-6 Multiplier used to measure the phase-angle difference between two equal frequencies.

meter face in degrees from a cosine table. That is, Eq. (11-4b) reduces to

$$V_{o\,dc} = \cos\theta \qquad \text{for } E_{x_p} = E_{y_p} = 4.47 \text{ V} \qquad (11\text{-}4c)$$

This point is explored further in Section 11.4.2.

Example 11-5

In Fig. 11-5, $E_{x_p} = E_{y_p} = 5$ V and the dc component of V_o is 1.25 V from Eq. (11-4b). Prove that there is $0°$ phase angle between E_x and E_y (since they are the same voltage).

Solution. From Eq. (11-4b),

$$\frac{20 \times 1.25}{5 \times 5} = \cos\theta = \frac{25}{25} = 1$$

Since $\cos 0° = 1$, $\theta = 0°$.

11-4.2   Phase-Angle Meter

Equation (11-4b) points the way to making a phase-angle meter. Assume that the peak values of E_x and E_y in Fig. 11-6(a) are scaled to 4.47 V by amplifiers or voltage dividers. Then a dc voltmeter is connected as shown in Fig. 4-6(a) to measure just the dc voltage component. The meter face then can be calibrated directly in degrees. The procedure is developed in Examples 11-6 and 11-7.

Example 11-6

The average value of V_o in Fig. 11-6(c) is 0, so the dc component of $V_o = 0$ V. Calculate θ.

Solution. From Eq. (11-4b), $\cos\theta = 0$, so $\theta = 90°$.

Example 11-7

Calculate $V_{o\,dc}$ for phase angles of (a) $\theta = 30°$; (b) $\theta = 45°$; (c) $\theta = 60°$.

Solution. From a trigonometric table, obtain the $\cos\theta$ and apply Eq. (11-4b).

θ (deg)	$\cos\theta$	$V_{o\,dc}(V)$
30	0.866	0.866
45	0.707	0.707
60	0.500	0.500
0	1.000	1.000
90	0.000	0.000

(The last two rows of this table come from Examples 11-5 and 11-6.)

The 0- to 1-V voltmeter scale can now be calibrated in degrees, 0 V for a $90°$ phase angle and 1.0 V for $0°$ phase angle. At 0.866 V, $\theta = 30°$, and so forth.

The phase angle meter does not indicate whether θ is a leading or lagging phase angle but only the phase difference between E_x and E_y.

11-4.3 Phase Angles Greater than 90°

The cosine of phase angles greater than $+90°$ or $-90°$ is a negative value. Therefore, V_o will be negative. This extends the capability of the phase angle meter in Example 11-7.

Example 11-8

Calculate $V_{o\,dc}$ for phase angles of (a) $\theta = \pm 90°$; (b) $\theta = \pm 120°$; (c) $\theta = \pm 135°$; (d) $\theta = \pm 150°$; (e) $\theta = \pm 180°$.

Solution. Using Eq. (11-4b) and tabulating results, we have

θ	$\pm 90°$	$\pm 120°$	$\pm 135°$	$\pm 150°$	$\pm 180°$
$V_{o\,dc}$	0 V	−0.5 V	−0.70 V	−0.866 V	−1 V

From the results of Examples 11-7 and 11-8, a ± 1-V voltmeter can be calibrated to read from 0 to 180°.

11-5 INTRODUCTION TO AMPLITUDE MODULATION

11-5.1 Need for Amplitude Modulation

Low-frequency audio or data signals cannot be transmitted from antennas of reasonable size. Audio signals can be transmitted by changing or *modulating* some characteristic of a higher-frequency *carrier* wave. If the amplitude of the carrier wave is changed in proportion to the audio signal, the process is called *amplitude modulation* (AM). Changing the frequency or the phase angle of the carrier wave results in *frequency modulation* (FM) and *phase-angle modulation* (PM), respectively.

Of course, the original audio signal must eventually be recovered by a process called *demodulation* or detection. The remainder of this section concentrates on using the multiplier for amplitude modulation.

11-5.2 Defining Amplitude Modulation

The introduction to amplitude modulation begins with the amplifier in Fig. 11-7(a). The input voltage E_c is amplified by a constant gain A. Amplifier output V_o is the product of gain A and E_c. Now suppose that the amplifier's gain is varied. This concept is represented by an arrow through A in Fig. 11-7(b). Assume that A is varied from 0 to a maximum and back to 0 as shown in Fig. 11-7(b) by the plot of A vs. t. This means that the amplifier multiplies the input voltage E_c by a different value (gain) over a period of time. V_o is now the amplitude of input E_c varied or multiplied by the amplitude of A. This process is an example of amplitude modulation, and the output voltage V_o is called the

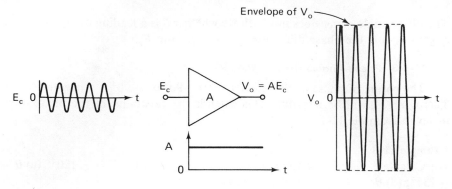

(a) Input E_c is amplified by constant gain A to give output $V_o = AE_c$

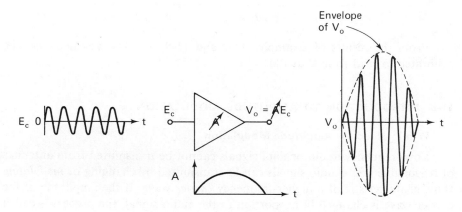

(b) If amplifier gain A is varied with time, the envelope of V_o is varied with time

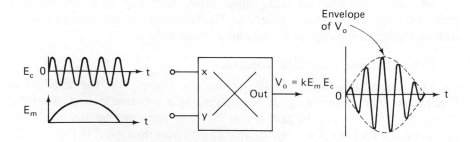

(c) If E_m varies as A in part (b), then V_o has the same general shape as in part (b)

Figure 11-7 Introduction to modulation.

amplitude modulated signal. Therefore, to obtain an amplitude-modulated signal (V_o), the amplitude of a high-frequency carrier signal (E_c) is varied by an intelligence or data signal A.

11-5.3 The Multiplier Used as a Modulator

From Section 11-5.2 and Fig. 11-7(b), V_o equals E_c multiplied by A. Therefore, amplitude modulation is a *multiplication process.* As shown in Fig. 11-7(c), E_c is applied to a multiplier's x input. E_m [having the same shape as A in Fig. 11-7(b)] is applied to the multiplier's y input. E_c is multiplied by a voltage that varies from 0 through a maximum and back to 0. So V_o has the same envelope as E_m. The multiplier can be considered a *voltage-controlled gain device* as well as an amplitude modulator. The waveshape shown is that of a *balanced modulator.* The reason for this name will be given in Section 11-6.3.

Note carefully in Fig. 11-7(c) that V_o is not a sine wave; that is, the peak values of successive half-cycles are different. This principle is used in Section 11-10 to show how a *frequency-shifter* (*heterodyne*) circuit works. But first, we examine amplitude modulation in greater detail.

11-5.4 Mathematics of the Balanced Modulator

A high-frequency sinusoidal *carrier wave* E_c is applied to one input of a multiplier. A lower-frequency audio or data signal is applied to the second input of a modulator and will be called the *modulating wave*, E_m. For test and analysis, both E_c and E_m will be sine waves described as follows.

Carrier wave, E_c:

$$E_c = E_{c_p} \sin 2\pi f_c t \qquad (11\text{-}5a)$$

where E_{c_p} is the peak value of the carrier wave and f_c is the carrier frequency.

Modulating wave, E_m:

$$E_m = E_{m_p} \sin 2\pi f_m t \qquad (11\text{-}5b)$$

where E_{m_p} is the peak value of the modulating wave and f_m is the modulating frequency.

Now let the carrier voltage E_c be applied to the x input of a multiplier as E_x, and let the modulating voltage E_m be applied to the y input of a multiplier as E_y. The multiplier's output voltage V_o is expressed as a product term from Eq. (11-1b) as

$$V_o = \frac{E_m E_c}{10} = \frac{E_{m_p} E_{c_p}}{10} (\sin 2\pi f_m t)(\sin 2\pi f_c t) \qquad (11\text{-}6)$$

Equation (11-6) is called the *product term*, because it represents the product of two sine waves with different frequencies. However, it is not in the form used by ham radio operators or communications personnel. They prefer the form obtained by applying to Eq. (11-6) the trigonometric identity

$$(\sin A)(\sin B) = \tfrac{1}{2}[\cos(A - B) - \cos(A + B)] \qquad (11\text{-}7)$$

Substituting Eq. (11-7) into Eq. (11-6), where $A = E_c$ and $B = E_m$, we have

$$V_o = \frac{E_{m_p}E_{c_p}}{20} \cos 2\pi(f_c - f_m)t - \frac{E_{m_p}E_{c_p}}{20} \cos 2\pi(f_c + f_m)t \qquad (11\text{-}8)$$

Equation (11-8) is analyzed in Section 11-5.5.

11-5.5 Sum and Difference Frequencies

Recall from Section 11-5.3 that E_c is a sine wave and E_m is a sine wave, but no part of V_o is a sine wave. V_o in Fig. 11-7(c) is expressed mathematically by either Eq. (11-6) or Eq. (11-8). But Eq. (11-8) shows that V_o is made up of two sine waves with frequencies different from either E_m or E_c. They are the *sum frequency* $f_c + f_m$ and the *difference frequency* $f_c - f_m$. The sum and difference frequencies are evaluated in Example 11-9.

Example 11-9

In Fig. 11-8, carrier signal E_c has a peak voltage of $E_{c_p} = 5$ V and a frequency of $f_c = 10{,}000$ Hz. The modulating signal E_m has a peak voltage of $E_{m_p} = 5$ V and a frequency of $f_m = 1000$ Hz. Calculate the peak voltage and frequency of (a) the sum frequency; (b) the difference frequency.

Solution. From Eq. (11-8), the peak value of both sum and difference voltages is

$$\frac{E_{m_p}E_{c_p}}{20} = \frac{5 \text{ V} \times 5 \text{ V}}{20} = 1.25 \text{ V}$$

The sum frequency is $f_c + f_m = 10{,}000$ Hz $+ 1000$ Hz $= 11{,}000$ Hz; the difference frequency is $f_c - f_m = 10{,}000$ Hz $- 1000$ Hz $= 9000$ Hz. Thus V_o is made up of the difference of two cosine waves:

$$V_o = 1.25 \cos 2\pi 9000t - 1.25 \cos 2\pi 11{,}000t$$

This result can be verified by connecting a wave or spectrum analyzer to the multiplier's output; a 1.25-V deflection occurs at 11,000 Hz and at 9000 Hz.

A cathode-ray oscilloscope can be used to show input and output voltages of the multiplier of Example 11-9. The product term for V_o is found from Eq. (11-6):

$$V_o = 2.5 \text{ V}\underbrace{(\sin 2\pi 10{,}000t)}_{E_c}\underbrace{(\sin 2\pi 1000t)}_{E_m}$$
$$= 2.5 \times \qquad E_c \qquad \times \qquad E_m$$

V_o is shown with E_m in the top drawing and with E_c in the bottom drawing of

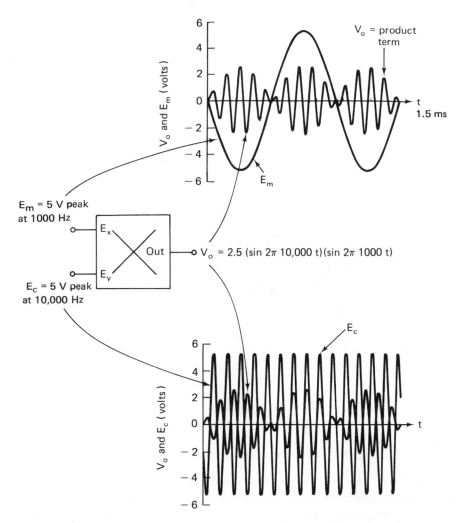

Figure 11-8 The multiplier as a balanced modulator.

Fig. 11-8. Observe that E_m and E_c have peak voltages of 5 V. The peak value of V_o is 2.5 V.

11-5.6 Side Frequencies and Side Bands

Another way of displaying the output of a modulator is by a graph showing the peak amplitude as a vertical line for each frequency. The resulting *frequency spectrum* is shown in Fig. 11-9(a). The sum and difference frequencies in V_o are called *upper* and *lower side* frequencies because they are above and below the carrier frequency on the graph.

When more than one modulating signal is applied to the modulator (y input) input in Fig. 11-8, each generates a sum and difference frequency in the

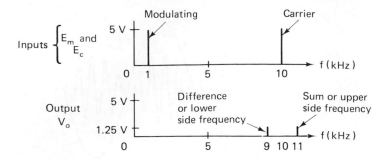

(a) Frequency spectrum for f_c = 10 kHz and f_m = 1 kHz in Example 11.8

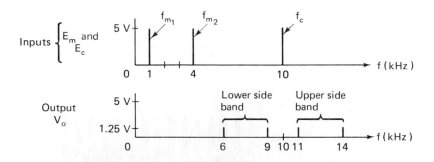

(b) Frequency spectrum for f_c = 10 kHz and f_{m_1} = 1 kHz, f_{m_2} = 4 kHz

Figure 11-9   Frequency spectrum for a balanced modulator.

output. Thus, there will be two side frequencies for each y input frequency, placed symmetrically on either side of the carrier. If the expected range of modulating frequencies is known, the resulting range of side frequencies can be predicted. For example, if the modulating frequencies range between 1 and 4 kHz, the lower side frequencies fall in a band between $(10 - 4)$ kHz = 6 kHz and $(10 - 1)$ kHz = 9 kHz. The band between 6 and 9 kHz is called the *lower side band*. For this same example, the *upper side band* ranges from $(10 + 1)$ kHz = 11 kHz to $(10 + 4)$ kHz = 14 kHz. Both upper and lower side bands are shown in Fig. 11-9(b).

11-6 STANDARD AMPLITUDE MODULATION

11-6.1 Amplitude Modulator Circuit

The circuits of Section 11-5 multiplied the carrier and modulating signals to generate a balanced output that is expressed either as (1) a product term, or (2) a sum and difference frequency. The classical or standard amplitude modu-

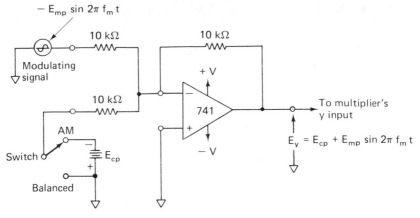

(a) Adder circuit to add carrier signal

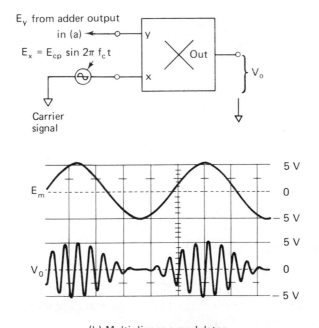

(b) Multiplier as a modulator

Figure 11-10 Circuit to demonstrate amplitude modulation or balanced modulation (see also Fig. 11-12).

lator (AM) adds the carrier term to the output. One way of adding the carrier term to generate an AM output is shown in Fig. 11-10(a). The modulating signal is fed into one input of an adder. A dc voltage equal to the peak value of the carrier voltage E_{c_p} is fed into the other input. The output of the adder is then fed into the y input of a multiplier, as shown in Fig. 11-10(b). The carrier signal is fed into the x input. The multiplier multiplies E_x by E_y, and its output voltage is the standard AM voltage given by either of the following equations.

$$V_o = \begin{cases} \dfrac{E_{c_p}^2}{10} \sin 2\pi f_c t \quad \text{(carrier term)} \\[2mm] + \\[2mm] \dfrac{E_{c_p} E_{m_p}}{10} (\sin 2\pi f_c t)(\sin 2\pi f_m t) \quad \text{(product term)} \end{cases} \tag{11-9}$$

or

$$V_o = \begin{cases} \dfrac{E_{c_p}^2}{10} \sin 2\pi f_c t \quad \text{(carrier term)} \\[2mm] + \\[2mm] \dfrac{E_{c_p} E_{m_p}}{20} \cos 2\pi(f_c - f_m)t \quad \text{(lower side frequency)} \\[2mm] - \\[2mm] \dfrac{E_{c_p} E_{m_p}}{20} \cos 2\pi(f_c + f_m)t \quad \text{(upper side frequency)} \end{cases} \tag{11-10}$$

The output voltage V_o is shown in Fig. 11-10(b). The voltage levels are worked out in the following example.

Example 11-10

In Fig. 11-10, $E_{c_p} = E_{m_p} = 5$ V. The carrier frequency $f_c = 10$ kHz, and the modulating frequency is $f_m = 1$ kHz. Evaluate the peak amplitudes of the output carrier and product terms.

Solution. From Eq. (11-9), the carrier term peak voltage is

$$\frac{(5 \text{ V})(5 \text{ V})}{10} = 2.5 \text{ V}$$

The product term peak voltage is

$$\frac{(5 \text{ V})(5 \text{ V})}{10} = 2.5 \text{ V}$$

The waveshape of V_o is shown in Fig. 11-10(b). Observe that the envelope of V_o is the same shape as E_m. This is characteristic of a standard amplitude modulator, AM, *not* of the balanced modulator.

11-6.2   Frequency Spectrum of a Standard AM Modulator

The signal frequencies present in V_o for the standard AM output of Fig. 11-10 are found from Eq. (11-10). Using the voltage values in Example 11-10, we have

$$\text{Carrier term} = 2.5\text{-V peak at } 10{,}000 \text{ Hz}$$
$$\text{Lower side frequency} = 1.25\text{-V peak at } 9000 \text{ Hz}$$
$$\text{Upper side frequency} = 1.25\text{-V peak at } 11{,}000 \text{ Hz}$$

These frequencies are plotted in Fig. 11-11 and should be compared with the balanced modulator of Fig. 11-9.

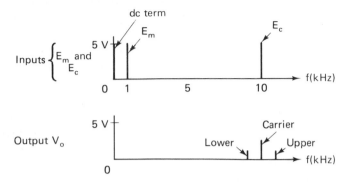

Figure 11-11 Frequency spectrum for a standard AM modulator, f_c = 10 kHz, f_m = 1 kHz.

11-6.3 Comparison of Standard AM Modulators and Balanced Modulators

If the switch in Fig. 11-10(a) is positioned to AM, V_o will contain three frequencies, f_c, $f_c + f_m$, and $f_c - f_m$. Observe that the envelope of V_o has the same shape as the intelligence signal E_m. This observation can be used to recover E_m from the AM signal.

If the switch in Fig. 11-10(a) is positioned to "Balanced," V_o will contain only the product term with only two frequencies $f_c + f_m$ and $f_c - f_m$. The envelope of V_o does *not* follow E_m. Since V_o does not contain f_c, this type of modulation is called *balanced modulation* in the sense that the carrier has been balanced out. It is also called *suppressed carrier modulation*, since the carrier is suppressed in the output.

For comparison of balanced and standard AM modulation, both outputs are shown together in Fig. 11-12.

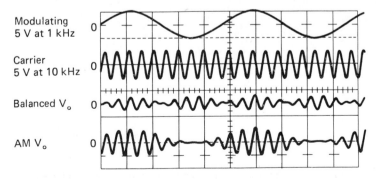

Figure 11-12 Comparison of balanced modulation and standard AM from Fig. 11-10.

11-7 DEMODULATING AN AM VOLTAGE

Demodulation, or *detection*, is the process of recovering a modulating signal E_m from the modulated output voltage V_o. To explain how this is accomplished, the AM modulated wave is applied to the y input of a multiplier as shown in Fig. 11-13. Each y input frequency is multiplied by the x input carrier frequency and generates a sum and difference frequency as shown in Fig. 11-13(b). Since only the 1-kHz frequency is the modulating signal, use a low-pass filter to extract E_m. Thus the demodulator is simply a multiplier with the carrier frequency applied to one input, and the AM signal to be demodulated is fed into the other input. The multiplier's output is fed into a low-pass filter whose output is the original modulating data signal E_m. Thus a multiplier plus a low-pass filter and carrier signal equals a demodulator.

Waveshapes at inputs and outputs of both the AM modulator and demodu-

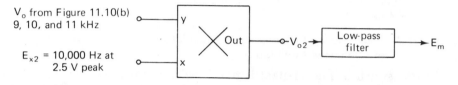

(a) Multiplier used as a demodulator

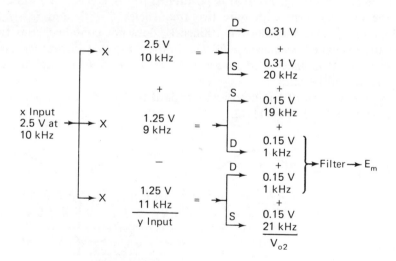

(b) Frequency and peak amplitude of signal components
at x-input, y-input, multiplier output and filter

Figure 11-13 The demodulator is a multiplier plus a low-pass filter.

lator are shown in Fig. 11-14. Note the unusual shape of V_{o2} because it contains six components [detailed in Fig. 11-13(b)].

11-8 DEMODULATING A BALANCED MODULATOR VOLTAGE

Modulating signal E_m is recovered from a balanced modulator by means of the same technique employed in Fig. 11-13 and Section 11-7. The only difference is due to the absent carrier frequency of 10 kHz at the demodulator's y input. This missing 10 kHz also eliminates both the dc and 20-kHz term in V_{o2}. The circuit arrangement of Fig. 11-15 was built to demonstrate the demodulating technique and show the resulting waveshapes. The demodulated E_m is not a pure sine wave, because only a simple filter was used. If f_c is increased to 100 kHz, E_m will be closer to being a pure sine wave.

The carrier's frequency fed into the demodulator should be *exactly* equal to the carrier's frequency driving the modulator.

11-9 SINGLE-SIDE-BAND MODULATION AND DEMODULATION

In the balanced modulator of Figs. 11-8 and 11-9, we could add a high-pass filter (see Chapter 12) to the modulator's output. If the filter removed all the lower side frequencies, the output is *single side band* (SSB). If the filter only attenuated the lower side frequencies (to leave a *vestige* of the lower side band), we would have a *vestigial* side-band modulator.

Assume that only one modulating frequency f_m is applied to our single-side-band modulator together with carrier f_c. Its output would be a single upper side frequency $f_c + f_m$. To demodulate this signal and recover f_m, all we have to do is connect the SSB signal $f_c + f_m$ to one multiplier input and f_c to the other input. According to the principles set forth in Section 11-5.4, the demodulators output would have a sum frequency of $(f_c + f_m) + f_c$ and a difference frequency of $(f_c + f_m) - f_c = f_m$. A low-pass filter would recover the modulating signal f_m and easily eliminate the high-frequency signal, whose frequency is $2f_c + f_m$.

11-10 FREQUENCY SHIFTING

In radio communication circuits, it is often necessary to shift a carrier frequency f_c with its accompanying side frequencies down to a lower intermediate frequency f_{IF}. This shift of each frequency is accomplished with the multiplier connections of Fig. 11-16(a). The modulated carrier signals are applied to the y input. A local oscillator is adjusted to a frequency, f_o, equal to the sum of the

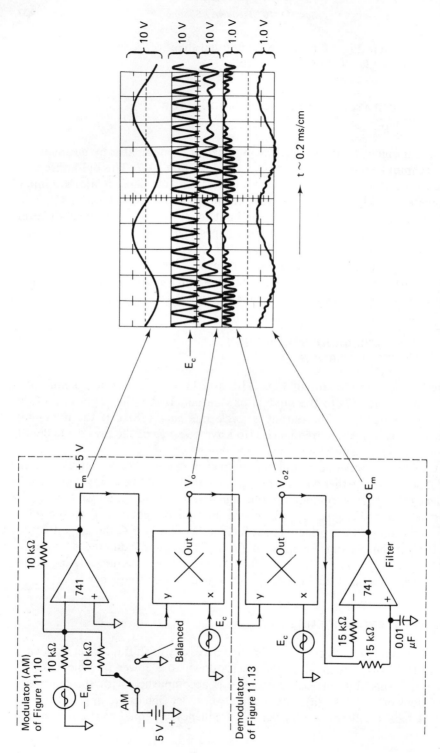

Figure 11-14 Voltage waveforms in an amplitude modulator and demodulator ($f_c = 10$ kHz, $f_m = 1$ kHz).

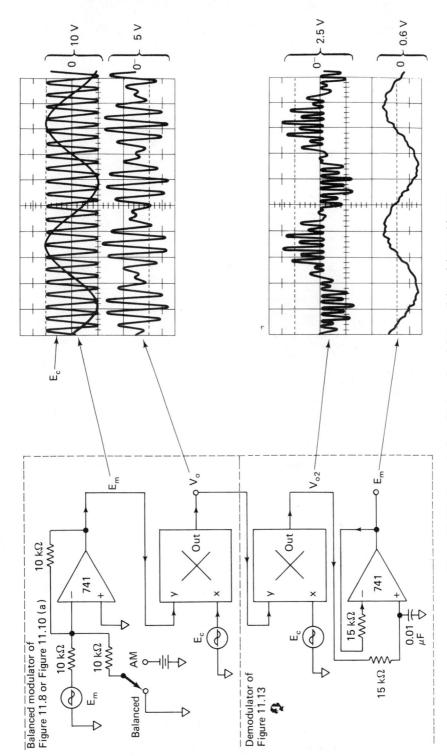

Figure 11-15 Demonstration of balanced modulator and demodulator with wave-shapes.

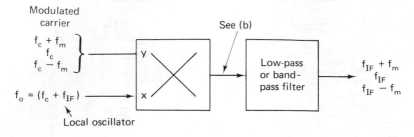

(a) Circuit for a frequency shifter

Frequency at y input (kHz)	Multiplier output	
	Peak (volts)	Frequency (kHz)
1005	$\dfrac{1 \times 5}{20} = 0.25$ V	$1455 + 1005 = 2460$ $1455 - 1005 = 450$
1000	$\dfrac{4 \times 5}{20} = 1.0$ V	$1455 + 1000 = 2455$ $1455 - 1000 = 455$
995	$\dfrac{1 \times 5}{20} = 0.25$ V	$1455 + 995 = 2450$ $1455 - 995 = 460$

(b) Frequencies present in multiplier output

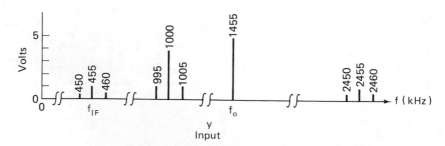

(c) y Input frequencies are shifted to the intermediate frequency

Figure 11-16 The multiplier as a frequency shifter.

carrier and desired intermediate frequency and applied to the x input. The frequencies present in the output of the multiplier are calculated in the following example.

Example 11-11

In Fig. 11-16(a), amplitudes and frequencies at each input are as follows: *y input:*

Peak amplitude (V)	Frequency (kHz)
1	$(f_c + f_m) = 1005$
4	$f_c = 1000$
1	$(f_c - f_m) = 995$

where f_c is the broadcasting station's carrier frequency and $(f_c + f_m)$ and $(f_c - f_m)$ are the upper and lower side frequencies due to a 5-kHz modulating frequency.

x input: The local oscillator is set for a 5-V peak sine wave at 1445 kHz, because the desired intermediate frequency is 455 kHz. Find the peak value and frequency of each signal component in the output of the multiplier.

Solution. From Eq. (11-10), the peak amplitude of each y input frequency is multiplied by the peak amplitude of the local oscillator frequency. This product is multiplied by $\frac{1}{20}$ ($\frac{1}{10}$ for the scale factor $\times$ $\frac{1}{2}$ from the trigonometric identity) to obtain the peak amplitude of the resulting sum and difference frequencies at the multiplier's output. The results are tabulated in Fig. 11-16(a).

All frequencies present in the multiplier's output are plotted on the frequency spectrum of Fig. 11-16(c). A low-pass filter or band-pass filter is used to pass only the three lower intermediate frequencies of 450, 455, and 460 kHz. The upper intermediate frequencies of 2450, 2455, and 2460 kHz may be used if desired, but they are usually filtered out.

We conclude from Example 11-11 that each frequency present at the y input is shifted down and up to new intermediate frequencies. The lower set of intermediate frequencies can be extracted by a filter. Thus, the information contained in the carrier f_c has been preserved and shifted to another subcarrier or intermediate frequency. The process of frequency shifting is also called *heterodyne*. The heterodyne principle will be used in Section 11-13 to construct a universal AM receiver that will demodulate standard AM, balance modulator and single-side-band signals.

11-11 ANALOG DIVIDER

An analog divider will give the ratio of two signals or provide gain control. It is constructed as shown in Fig. 11-17 by inserting a multiplier in the feedback loop of an op amp. Since the op amp's $(-)$ input draws negligible current, the

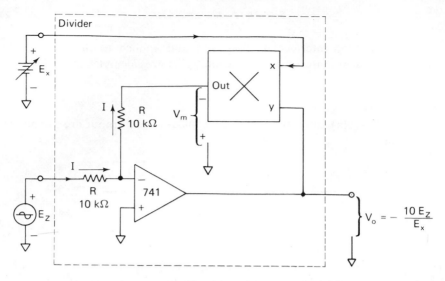

Figure 11-17 Division with an op amp and a multiplier.

current I is equal in the equal resistors R. Therefore, the output voltage of the multiplier V_m is equal in magnitude but opposite in polarity (with respect to ground) to E_z or

$$E_z = -V_m \tag{11-11a}$$

But V_m is also equal to one-tenth (scale factor) of the product of input E_x and output of the op amp V_o. Substituting for V_m yields

$$E_z = -\frac{V_o E_x}{10} \tag{11-11b}$$

Solving for V_o, we obtain

$$V_o = -\frac{10 E_z}{E_x} \tag{11-11c}$$

Equation (11-11c) shows that the divider's output V_o is proportional to the ratio of inputs E_z and E_x. E_x should never be allowed to go to 0 V or to a negative voltage, because the op amp will saturate. E_z can be either positive, negative, or 0 V. Note that the divider can be viewed as a voltage gain $10/E_x$ acting on E_z. So if E_x is changed, the gain will change. This voltage control of the gain is useful in automatic gain-control circuits.

11-12 FINDING SQUARE ROOTS

A divider can be made to find square roots by connecting both inputs of the multiplier to output of the op amp (see Fig. 11-18). Equation (11-11a) also pertains to Fig. 11-18. But now V_m is one-tenth (scale factor) of $V_o \times V_o$ or

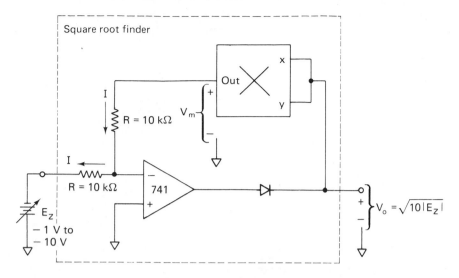

Figure 11-18 Square rooting with an op amp and a multiplier.

$$-E_z = V_m = \frac{V_o^2}{10} \tag{11-12a}$$

Solving for V_o (eliminate $\sqrt{-1}$) yields

$$V_o = \sqrt{10 \, |E_z|} \tag{11-12b}$$

Equation (11-12b) states that V_o equals the square root of 10 times the *magnitude* of E_z. E_z must be a negative voltage, or else the op amp saturates. The range of E_z is between -1 and -10 V. Voltages smaller than -1 V will cause inaccuracies. The diode prevents $(-)$ saturation for positive E_z.

11-13 UNIVERSAL AMPLITUDE MODULATION RECEIVER

11-13.1 Tuning and Mixing

The ordinary automobile or household AM radio can receive only standard AM signals that occupy the AM broadcast band, about 500 to 1500 kHz. This type of radio receiver cannot extract the audio or data signals from single-sideband (CB) or suppressed carrier transmission.

Figure 11-19 shows a receiver that will receive any type of AM transmission, carrier plus side bands, side bands without carrier, or a single side band (either upper or lower). To understand its operation, assume that a station is transmitting a 5-kHz audio signal that modulates a 1005-kHz carrier wave. The station transmits a standard AM frequency spectrum of the 1005-kHz carrier and both lower and upper side frequencies of 1000 and 1010 kHz [see Fig. 11-19(a)].

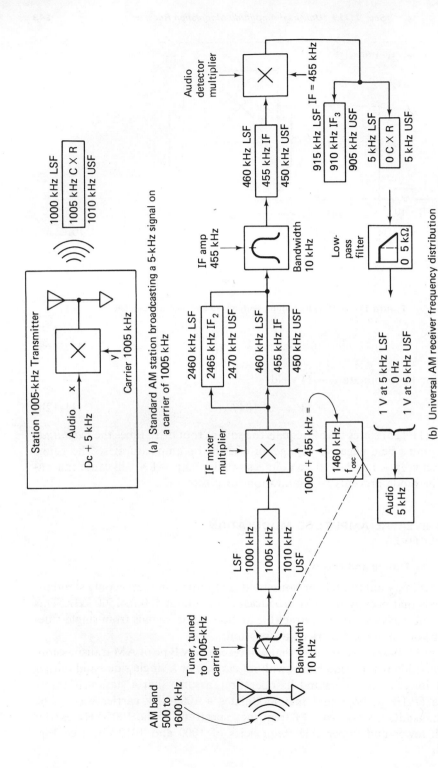

Figure 11-19 The superheterodyne receiver in (b) can demodulate or detect audio signals from the standard AM transmission in (a) and also single-side-band or suppressed carrier AM.

(a) Standard AM station broadcasting a 5-kHz signal on a carrier of 1005 kHz

(b) Universal AM receiver frequency distribution

In Fig. 11-19(b) the receiver's *tuner* is tuned to select this one station's 10-kHz band of frequencies out of the entire broadcast band of frequencies that are present on the receiver's antenna. A *local oscillator* in the radio is designed to produce a signal that *tracks* the tuner and is always 455 kHz higher than the tuner frequency. The oscillator and tuner output frequencies are multiplied by the *IF mixer*. The IF mixer acts as a frequency shifter to shift the incoming radio frequency carrier down to an *intermediate-frequency* (IF) carrier of 455 kHz.

11-13.2 Intermediate-Frequency Amplifier

The output of the IF mixer contains both sum frequencies (2465-kHz IF_2 carrier) and difference frequencies (455-kHz IF carrier). Only the difference frequencies are amplified by the tuned high-gain *IF amplifier*. This first frequency shift (heterodyning) is performed so that most of the signal amplification is done by a single narrow-band tuned-IF amplifier that usually has three stages of gain. *Any* station carrier that is selected by the tuner is shifted by the local oscillator and mixer multiplier down to the IF frequency for amplification. This frequency downshift scheme is used because it is much easier to build a reliable narrow-band IF amplifier (10-kHz bandwidth centered on a 455-kHz carrier) than it is to build an amplifier that can provide equal amplification of, *and* select 10-kHz bandwidths over, the *entire* AM broadcast band.

11-13.3 Detection Process

The output of the IF amplifier is multiplied by the IF frequency in the *audio detector multiplier*. The term *detection* means that we are going to *detect* or *demodulate* the audio signal from the 455-kHz IF carrier. The audio detector shifts the incoming IF carrier and side frequencies up and down as sum and difference frequencies. Only the difference frequencies are transmitted through the low-pass filter in Fig. 11-19(b). The astute reader will note that the *low-pass filter* output frequencies are *not* labeled (+) 5 kHz for the upper side frequency and (−) 5 kHz for the lower side frequency. If you work the mathematics out using sine waves for audio, carrier, local oscillator, and IF signals, it turns out that both audio 5-kHz signals are in phase (as negative cosine waves). The output of the low-pass filter is applied to an audio amplifier and finally to a speaker.

11-13.4 Universal AM Receiver

Why will this receiver do what most other receivers cannot do? The previous sections dealt with a standard AM transmission carrier plus *both* upper and lower side frequencies. Suppose that you eliminated the 1005-kHz carrier at the transmitter in Fig. 11-19(a). Note how the carrier is identified by enclosure in a rectangle as it progresses through the receiver in Fig. 11-19(b). If no carrier enters the receiver, but only the side frequencies, both audio signals (USF and LSF) will still enter the audio amplifier. Thus this receiver can recover audio information from either standard AM or balanced AM modulation. An AM car radio will *not* recover the audio signal from balanced AM transmissions.

Next suppose that you eliminated (by filters) the carrier *and* upper side frequency from the transmitter in Fig. 11-19(a); only the lower side frequency of 1000 kHz would be broadcasted. The entire lower side band would occupy 1000 to 1005 kHz. This is *single-side-band transmission*. At the receiver, the tuner would select 1000 kHz (to 1005 kHz). The IF amplifier would output 460 kHz (to 455 kHz). Finally, the low-pass filter would output 5 kHz (to 0 kHz); thus this receiver can also receive single-side-band transmission. The versatility of this type of receiver is inherent in the design. It requires no switches to activate circuit changes for different types of AM modulation.

PROBLEMS

11-1. Find V_o in Fig. 11-1 for the following combination of inputs: (a) $x = 5$ V, $y = 5$ V; (b) $x = -5$ V, $y = 5$ V; (c) $x = 5$ V, $y = -5$ V; (d) $x = -5$ V, $y = -5$ V.

11-2. State the operating point quadrant for each combination in Problem 11-1 [see Fig. 11-2(a)].

11-3. What is the name of the procedure used to make $V_o = 0$ V when both x and y inputs are at 0 V?

11-4. Find V_o in Fig. 11-4 if $E_i = -3$ V.

11-5. The peak value of E_i in Fig. 11-5 is 8 V, and its frequency is 400 Hz. Evaluate the output's (a) dc terms; (b) ac term.

11-6. In Fig. 11-6, $E_{x_p} = 10$ V, $E_{y_p} = 10$ V, and $\theta = 30°$. Find V_o.

11-7. Repeat Problem 11-6 for $\theta = -30°$.

11-8. In the balanced modulator of Fig. 11-8, E_x is a 15-kHz sine wave at 8 V peak and E_y is a 3-kHz sine wave at 5 V peak. Find the peak voltage of each frequency in the output.

11-9. In Fig. 11-8, the carrier frequency is 15 kHz. The modulating frequencies range between 1 and 2 kHz. Find the upper and lower side bands.

11-10. The switch is on AM in Fig. 11-10. The modulating frequency is 10 kHz at 5 V peak. The carrier is 100 kHz at 8 V peak. Identify the peak value and each frequency contained in the output.

11-11. If the switch is thrown to "Balanced" in Problem 11-10, what changes result in the output?

11-12. The x input of Fig. 11-13 is three sine waves of 5 V at 20 kHz, 2.0 V at 21 kHz, and 2.0 V at 19 kHz. The y input is 5 V at 20 kHz. What are the output signal frequency components?

11-13. It is desired to shift a 550-kHz signal to a 455-kHz intermediate frequency. What frequency should be generated by the local oscillator?

11-14. $E_x = 10$ V and $E_z = -1$ V in Fig. 11-17. Find V_o.

active filters

ıⁿⁱⁿı

12

12-0 INTRODUCTION

A *filter* is a circuit that is designed to pass a specified band of frequencies while attenuating all signals outside this band. Filter networks may be either active or passive. *Passive filter networks* contain only resistors, inductors, and capacitors. *Active filters*, which are the only type covered in this text, employ transistors or op amps plus resistors, inductors, and capacitors. Inductors are not often used in active filters, because they are bulky and costly and may have large internal resistive components.

There are four types of filters: *low-pass*, *high-pass*, *band-pass*, and *band-elimination* (also referred to as *band-reject* or *notch*) filters. Figure 12-1 illustrates frequency-response plots for the four types of filters. A low-pass filter is a circuit that has a constant output voltage from dc up to a *cutoff frequency* f_c. As the frequency increases above f_c, the output voltage is attenuated (decreases). Figure 12-1(a) is a plot of the magnitude of the output voltage of a low-pass filter versus frequency. The solid line is a plot for the ideal low-pass filter, while the dashed lines indicate the curves for practical low-pass filters. The range of frequencies that are *transmitted* is known as the *pass band*. The range of frequencies that are *attenuated* is known as the *stop band*. The cutoff frequency, f_c, is also called the 0.707 frequency, the 3-dB frequency, the corner frequency, or the break frequency.

High-pass filters attenuate the output voltage for all frequencies below the cutoff frequency f_c. Above f_c, the magnitude of the output voltage is constant.

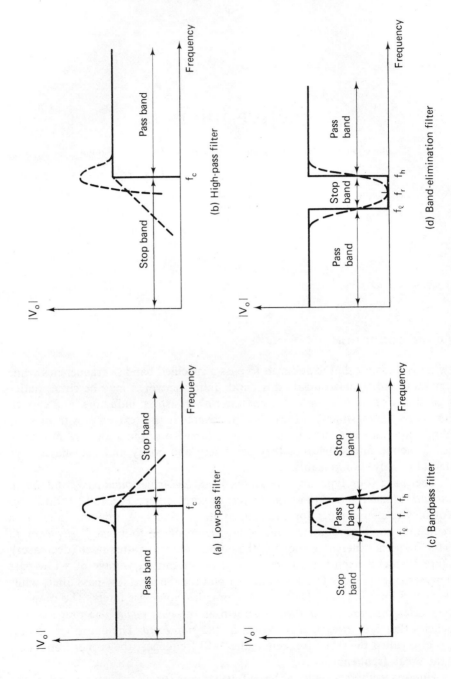

Figure 12-1 Frequency response for four categories of filters.

(a) Low-pass filter

(b) High-pass filter

(c) Bandpass filter

(d) Band-elimination filter

Figure 12-1(b) is the plot for ideal and practical high-pass filters. The solid line is the ideal curve, while the dashed curves show how practical high-pass filters deviate from the ideal.

Band-pass filters pass only a band of frequencies while attenuating all frequencies outside the band. Band-elimination filters perform in an exactly opposite way; that is, band-elimination filters reject a specified band of frequencies while passing all frequencies outside the band. Typical frequency-response plots for band-pass and band-elimination filters are shown in Fig. 12-1(c) and (d). Once again, the solid line represents the ideal plot, while dashed lines show the practical curves.

12-1   BASIC LOW-PASS FILTER

12-1.1   Introduction

The circuit of Fig. 12-2(a) is a commonly used low-pass active filter. The filtering is done by the RC network, and the op amp is used as a unity-gain amplifier. The resistor R_f is equal to R and is included for dc offset. [At dc, the capacitive reactance is infinite and the dc resistance path to ground for both input terminals should be equal (see Section 9-4).]

The differential voltage between pins 2 and 3 is essentially 0 V. Therefore, the voltage across capacitor C equals output voltage V_o, because this circuit is a voltage follower. E_i divides between R and C. The capacitor voltage equals V_o and is

$$V_o = \frac{1/j\omega C}{R + 1/j\omega C} \times E_i \qquad (12\text{-}1a)$$

where ω is the frequency of E_i in radians per second ($\omega = 2\pi f$) and j is equal to $\sqrt{-1}$. Rewriting Eq. (12-1a) to obtain the closed-loop voltage gain A_{CL}, we have

$$A_{CL} = \frac{V_o}{E_i} = \frac{1}{1 + j\omega RC} \qquad (12\text{-}1b)$$

To show that the circuit of Fig. 12-2(a) is a low-pass filter, consider how A_{CL} in Eq. (12-1b) varies as frequency is varied. At very low frequencies, that is, as ω approaches 0, $|A_{CL}| = 1$, and at very high frequencies, as ω approaches infinity, $|A_{CL}| = 0$. ($|\quad|$ means magnitude.)

Figure 12-2(b) is a plot of $|A_{CL}|$ versus ω and shows that for frequencies *greater* than the cutoff frequency ω_c, $|A_{CL}|$ decreases at a rate of 20 dB/decade. This is the same as saying that the voltage gain is divided by 10 when the frequency of ω is increased by 10.

12-1.2   Designing the Filter

The cutoff frequency ω_c is defined as that frequency of E_i where $|A_{CL}|$ is reduced to 0.707 times its low-frequency value. This important point will be

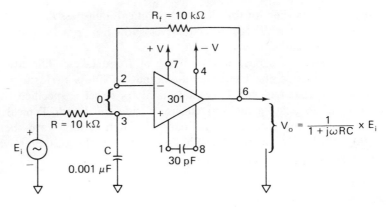

(a) Low-pass filter for a roll-off of − 20 db/decade

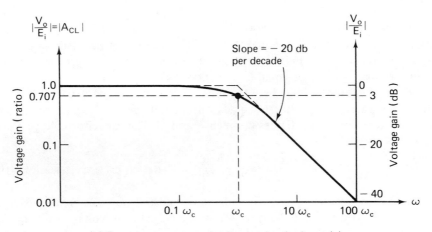

(b) Frequency-response plot for the circuit of part (a)

Figure 12-2 Low-pass filter and frequency-response plot for a filter with a −20-dB/decade roll-off.

discussed further in Section 12-1.3. The cutoff frequency is evaluated from

$$\omega_c = \frac{1}{RC} = 2\pi f_c \qquad (12\text{-}2a)$$

where ω_c is the cutoff frequency in radians per second, f_c is the cutoff frequency in hertz, R is in ohms, and C is in farads. Equation (12-2a) may be rearranged to solve for C:

$$C = \frac{1}{\omega_c R} = \frac{1}{2\pi f_c R} \qquad (12\text{-}2b)$$

Example 12-1

Let $R = 10 \text{ k}\Omega$ and $C = 0.001 \ \mu\text{F}$ in Fig. 12-2(a); what is the cutoff frequency?

Solution. By Eq. (12-2a),

$$\omega_c = \frac{1}{(10 \times 10^3)(0.001 \times 10^{-6})} = 100 \text{ krad/s}$$

or

$$f_c = \frac{\omega_c}{6.28} = \frac{100 \times 10^3}{6.28} = 15.9 \text{ kHz}$$

Example 12-2

For the low-pass filter of Fig. 12-2(a), calculate C for a cutoff frequency of 2 kHz and $R = 10$ kΩ.

Solution. From Eq. (12-2b),

$$C = \frac{1}{\omega_c R} = \frac{1}{(6.28)(2 \times 10^3)(10 \times 10^3)} \cong 0.008 \ \mu\text{F}$$

Example 12-3

Calculate C for Fig. 12-2(a) for a cutoff frequency of 30 krad/s and $R = 10$ kΩ.

Solution. From Eq. (12-2b),

$$C = \frac{1}{\omega_c R} = \frac{1}{(30 \times 10^3)(10 \times 10^3)} = 0.0033 \ \mu\text{F}$$

Conclusion

The design of a low-pass filter similar to Fig. 12-2(a) is accomplished in three steps:

1. Choose the cutoff frequency—either ω_c or f_c.
2. Choose the input resistance R, usually between 10 and 100 kΩ.
3. Calculate C from Eq. (12-2b).

12-1.3 Filter Response

The value of A_{CL} at ω_c is found by letting $\omega RC = 1$ in Eq. (12-1b):

$$A_{CL} = \frac{1}{1 + j1} = \frac{1}{\sqrt{2} \angle 45°} = 0.707 \angle -45°$$

Therefore, the magnitude of A_{CL} at ω_c is

$$|A_{CL}| = \frac{1}{\sqrt{2}} = 0.707 = -3 \text{ dB}$$

and the phase angle is $-45°$.

The solid curve in Fig. 12-2(b) shows how the magnitude of the actual frequency response deviates from the straight dashed-line approximation in the vicinity of ω_c. At $0.1\omega_c$, $|A_{CL}| = 1$ (0 dB), and at $10\omega_c$, $|A_{CL}| = 0.1$ (−20 dB). Table 12-1 gives both the magnitude and the phase angle for different values of ω between $0.1\,\omega_c$ and $10\omega_c$.

Table 12-1 Magnitude and phase angle for the low-pass filter of Fig. 12-2(a)

ω	$|A_{CL}|$	Phase angle (deg)
$0.1\omega_c$	1.0	-6
$0.25\omega_c$	0.97	-14
$0.5\omega_c$	0.89	-27
ω_c	0.707	-45
$2\omega_c$	0.445	-63
$4\omega_c$	0.25	-76
$10\omega_c$	0.1	-84

Many applications require steeper roll-offs after the cutoff frequency. One common filter configuration that gives steeper roll-offs is the *Butterworth filter*.

12-2 INTRODUCTION TO THE BUTTERWORTH FILTER

In many low-pass filter applications, it is necessary for the closed-loop gain to be as close to 1 as possible within the pass band. The *Butterworth filter* is best suited for this type of application. The Butterworth filter is also called a *maximally flat* or *flat-flat* filter, and all filters in this chapter will be of the Butterworth type. Figure 12-3 shows the ideal (solid line) and the practical (dashed lines)

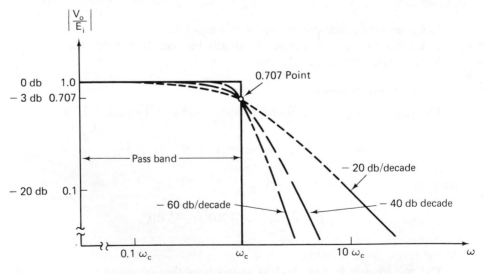

Figure 12-3 Frequency-response plots for three types of low-pass Butterworth filters.

frequency response for three types of Butterworth filters. As the roll-offs become steeper, they approach the ideal filter more closely.

Two active filters similar to Fig. 12-2(a) could be coupled together to give a roll-off of −40 dB/decade. This would not be the most economical design, because it would require two op amps. In Section 12-3.1, it is shown how one op amp can be used to build a Butterworth filter with a single op amp to give a −40-dB/decade roll-off. Then in Section 12-4, a −40-dB/decade filter will be cascaded with a −20-dB/decade filter to produce a −60-dB/decade filter.

Butterworth filters are not designed to keep a constant phase angle at the cutoff frequency. A basic low-pass filter of −20-dB/decade has a phase angle of −45° at ω_c. A −40-dB/decade Butterworth filter has a phase angle of −90° at ω_c and a −60-dB/decade filter has a phase angle of −135° at ω_c. Therefore, for each increase of −20-dB/decade, the phase angle will increase by −45°. We now proceed to a Butterworth filter that has a roll-off steeper than −20 dB/decade.

12-3 −40-DB/DECADE LOW-PASS BUTTERWORTH FILTER

12-3.1 Simplified Design Procedure

The circuit of Fig. 12-4(a) is one of the most commonly used low-pass filters. It produces a roll-off of −40 dB/decade; that is, after the cut-off frequency, the magnitude of A_{CL} decreases by 40 dB as ω increases to $10\omega_c$. The solid line in Fig. 12-4(b) shows the actual frequency-response plot, which is explained in more detail in Section 12-3.2.

The op amp is connected for dc unity gain. Resistor R_f is included for dc offset, as explained in Section 9-4.

Since the op amp circuit is basically a voltage follower (unity-gain amplifier), the voltage across C_1 equals output voltage, V_o.

The design of the low-pass filter of Fig. 12-4(a) is greatly simplified by making resistors R_1 and R_2 equal. There are only four steps in the design procedure:

1. Choose the cutoff frequency, ω_c or f_c.
2. Let $R_1 = R_2 = R$, and choose a convenient value between 10 and 100 kΩ. Choose $R_f = 2R$.
3. Calculate C_1 from

$$C_1 = \frac{0.707}{\omega_c R} \tag{12-3}$$

4. Choose

$$C_2 = 2C_1 \tag{12-4}$$

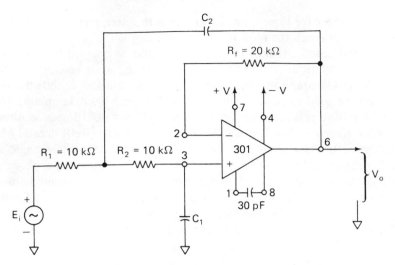

(a) Low-pass filter for a roll-off of − 40 db/decade

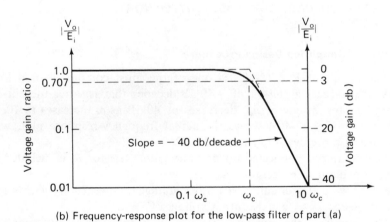

(b) Frequency-response plot for the low-pass filter of part (a)

Figure 12-4 Circuit and frequency plot for a low-pass filter of −40 dB/decade.

Example 12-4

Determine (a) C_1 and (b) C_2 in Fig. 12-4(a) for a cutoff frequency of 30 krad/s. Let $R_1 = R_2 = R = 10$ kΩ.

Solution. (a) From Eq. (12-3),

$$C_1 = \frac{0.707}{(30 \times 10^3)(10 \times 10^3)} = 0.0024 \ \mu F$$

(b) $C_2 = 2C_1 = 2(0.0024 \ \mu F) = 0.0048 \ \mu F.$

12-3.2 Filter Response

The dashed curve in Fig. 12-4(b) shows that the filter of Fig. 12-4(a) not only has a steeper roll-off after ω_c than does Fig. 12-2(a) but also remains at 0 dB almost up to about $0.25\omega_c$. The phase angles for the circuit of Fig. 12-4(a) range from $0°$ at $\omega = 0$ rad/s (dc condition) to $-180°$ as ω approaches ∞ (infinity). Table 12-2 compares magnitude and phase angle for the low-pass filters of Figs. 12-2(a) and 12-4(a) from $0.1\omega_c$ to $10\omega_c$.

Table 12-2 Magnitude and phase angle for Figs. 12-2(a) and 12-4(a)

ω	$|A_{CL}|$ Fig. 12-2(a)	$|A_{CL}|$ Fig. 12-4(a)	Phase angle (deg) Fig. 12-2(a)	Phase angle (deg) Fig. 12-4(a)
$0.1\omega_c$	1.0	1.0	-6	-8
$0.25\omega_c$	0.97	0.998	-14	-21
$0.5\omega_c$	0.89	0.97	-27	-43
ω_c	0.707	0.707	-45	-90
$2\omega_c$	0.445	0.24	-63	-137
$4\omega_c$	0.25	0.053	-76	-143
$10\omega_c$	0.1	0.01	-84	-172

The next low-pass filter cascades the filter of Fig. 12-2(a) with the filter of Fig. 12-4(a) to form a roll-off of -60 dB/decade.

As will be shown, the capacitors are the only values that have to be calculated.

12-4 −60-DB/DECADE LOW-PASS BUTTERWORTH FILTER

12-4.1 Simplified Design Procedure

The low-pass filter of Fig. 12-5(a) is built using one low-pass filter of -40 dB/decade cascaded with another of -20 dB/decade to give an overall roll-off of -60 dB/decade. The overall closed-loop gain A_{CL} is the gain of the first filter times the gain of the second filter, or

$$A_{CL} = \frac{V_o}{E_i} = \frac{V_{o1}}{E_i} \times \frac{V_o}{V_{o1}} \tag{12-5}$$

For a Butterworth filter, the magnitude of A_{CL} must be 0.707 at ω_c; to guarantee that the frequency response is flat in the pass band, use the following design steps:

1. Choose the cutoff frequency, ω_c or f_c.
2. Choose the input resistors to be equal ($R_1 = R_2 = R_3 = R$); values between 10 and 100 kΩ are typical.

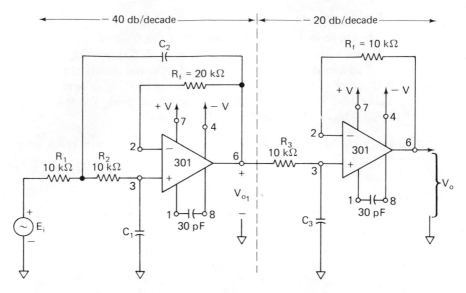

(a) Low-pass filter for a roll off of − 60 db/decade

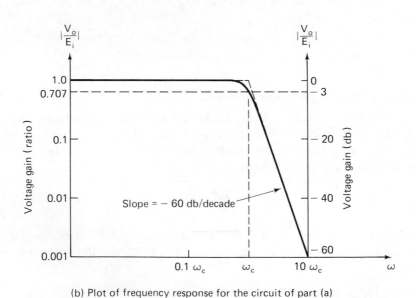

(b) Plot of frequency response for the circuit of part (a)

Figure 12-5 Low-pass filter designed for a roll-off of −60 dB/decade and corresponding frequency-response plot.

3. Calculate C_3 from Eq. (12-2b), which is rewritten as

$$C_3 = \frac{1}{\omega_c R} \tag{12-6}$$

4. $C_1 = \frac{1}{2}C_3.$ (12-7)
5. $C_2 = 2C_3.$ (12-8)

Example 12-5

For the low-pass filter of Fig. 12-5(a), calculate (a) C_3, (b) C_1, and (c) C_2 for a cutoff frequency of 30 krad/s. $R_1 = R_2 = R_3 = 10 \ k\Omega$.
Solution. (a) From Eq. (12-6),

$$C_3 = \frac{1}{(30 \times 10^3)(10 \times 10^3)} = 0.0033 \ \mu F$$

(b) $C_1 = \frac{1}{2}C_3 = 0.0017 \ \mu F.$
(c) $C_2 = 2C_3 = 0.0066 \ \mu F.$

Example 12-5 shows that the capacitors C_1 and C_2 of Fig. 12-5(a) are different from those of Fig. 12-4(a), although the cutoff frequency is the same. This is necessary so that $|A_{CL}|$ remains at 0 dB in the pass band until the cutoff frequency is nearly reached; then $|A_{CL}| = 0.707$ at ω_c.

12-4.2 Filter Response

The solid line in Fig. 12-5(b) is the actual plot of the frequency response for Fig. 12-5(a). The dashed curve in the vicinity shows the straight-line approximation. Table 12-3 compares the magnitudes of A_{CL} for the three low-pass filters presented in this chapter. Note that the $|A_{CL}|$ for Fig. 12-5(a) remains quite close to 1 (0 dB) until the cutoff frequency, ω_c; then the steep roll-off occurs.

Table 12-3 $|A_{CL}|$ for the low-pass filters of Figs. 12-2(a), 12-4(a), and 12-5(a)

ω	−20 dB/decade; Fig. 12-2(a)	−40 dB/decade; Fig. 12-4(a)	−60 dB/decade; Fig. 12-5(a)
$0.1\omega_c$	1.0	1.0	1.0
$0.25\omega_c$	0.97	0.998	0.999
$0.5\omega_c$	0.89	0.97	0.992
ω_c	0.707	0.707	0.707
$2\omega_c$	0.445	0.24	0.124
$4\omega_c$	0.25	0.053	0.022
$10\omega_c$	0.1	0.01	0.001

The phase angles for the low-pass filter of Fig. 12-5(a) range from 0° at $\omega = 0$ (dB condition) to $-270°$ as ω approaches ∞. Table 12-4 compares the phase angles for the three low-pass filters.

Table 12-4   Phase angles for the low-pass filters of Figs. 12-2(a), 12-4(a), and 12-5(a)

ω	$-20\ dB/decade;$ Fig. 12-2(a)	$-40\ dB/decade;$ Fig. 12-4(a)	$-60\ dB/decade;$ Fig. 12-5(a)
$0.1\omega_c$	$-6°$	$-8°$	$-12°$
$0.25\omega_c$	$-14°$	$-21°$	$-29°$
$0.5\omega_c$	$-27°$	$-43°$	$-60°$
ω_c	$-45°$	$-90°$	$-135°$
$2\omega_c$	$-63°$	$-137°$	$-210°$
$4\omega_c$	$-76°$	$-143°$	$-226°$
$10\omega_c$	$-84°$	$-172°$	$-256°$

12-5   HIGH-PASS BUTTERWORTH FILTERS

12-5.1   Introduction

A high-pass filter is a circuit that attenuates all signals below a specified cutoff frequency ω_c and passes all signals whose frequency is above the cutoff frequency. Thus a high-pass filter performs the opposite function of the low-pass filter.

Figure 12-6 is a plot of the magnitude of the closed-loop gain versus ω for

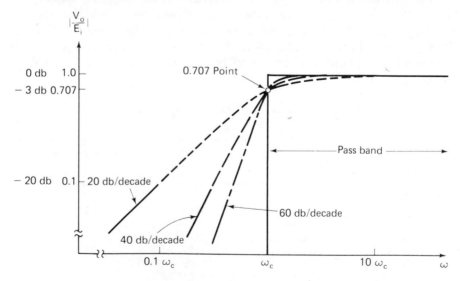

Figure 12-6   Comparison of frequency response for three high-pass Butterworth filters.

three types of Butterworth filters. The phase angle for a circuit of 20 dB/decade is $+45°$ at ω_c. Phase angles at ω_c increase by $+45°$ for each increase of 20 dB/decade. The phase angles for these three types of high-pass filters are compared in Section 12-5.5.

In this text, the design of high-pass filters will be similar to that of the low-pass filters. In fact, the only difference will be the position of the filtering capacitors and resistors.

12-5.2 20-dB/Decade Filter

Compare the high-pass filter of Fig. 12-7(a) with the low-pass filter of Fig. 12-2(a) and note that C and R are interchanged. The feedback resistor R_f is

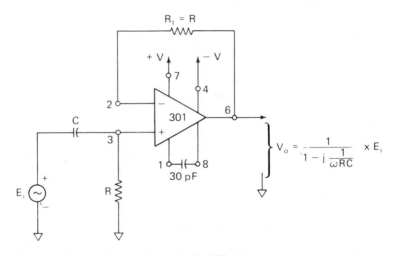

(a) High-pass filter with a roll-off of 20 db/decade

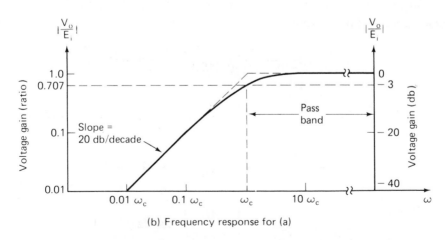

(b) Frequency response for (a)

Figure 12-7 Basic high-pass filter, 20 dB/decade.

included to minimize dc offset. Since the op amp is connected as a unity-gain follower in Fig. 12-7(a), the output voltage V_o equals the voltage across R and is expressed by

$$V_o = \frac{1}{1 - j(1/\omega RC)} \times E_i \tag{12-9}$$

When ω approaches 0 rad/s in Eq. (12-9), V_o approaches 0 V. At high frequencies, as ω approaches infinity, V_o equals E_i. Since the circuit is not an ideal filter, the frequency response is not ideal, as shown by Fig. 12-7(b). The solid line is the actual response, while the dashed lines show the straight-line approximation. The magnitude of the closed-loop gain equals 0.707 when $\omega RC = 1$. Therefore, the cutoff frequency ω_c is given by

$$\omega_c = \frac{1}{RC} = 2\pi f_c \tag{12-10a}$$

or

$$R = \frac{1}{\omega_c C} = \frac{1}{2\pi f_c C} \tag{12-10b}$$

The reason for solving for R and not C in Eq. (12-10b) is that for high-pass filters, usually C is chosen along with ω_c and R is calculated. The steps needed in designing Fig. 12-7(a) are

1. Choose the cutoff frequency, ω_c or f_c.
2. Choose a convenient value of C.
3. Calculate R from Eq. (12-10b).
4. Choose $R_f = R$.

Example 12-6
   Calculate R in Fig. 12-7(a) if $C = 0.002\ \mu F$ and $f_c = 10$ kHz.
Solution. From Eq. (12-10b),

$$R = \frac{1}{(6.28)(10 \times 10^3)(0.002 \times 10^{-6})} = 8\ k\Omega$$

Example 12-7
   In Fig. 12-7(a) if $R = 22\ k\Omega$ and $C = 0.01\ \mu F$, calculate (a) ω_c and (b) f_c.
Solution. (a) From Eq. (12-10a),

$$\omega_c = \frac{1}{(22 \times 10^3)(0.01 \times 10^{-6})} = 4.54\ krad/s$$

(b)

$$f_c = \frac{\omega_c}{2\pi} = \frac{4.54 \times 10^3}{6.28} = 724\ Hz$$

12-5.3 40-dB/Decade Filter

The circuit of Fig. 12-8(a) is to be designed as a high-pass Butterworth filter with a roll-off of 40 dB/decade below the cutoff frequency, ω_c. To satisfy the Butterworth criteria, the frequency response must be 0.707 at ω_c and be 0 dB in the pass band. These conditions will be met if the following design procedure is followed:

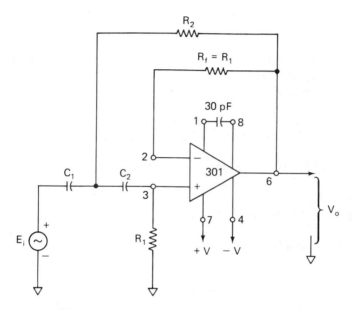

(a) High-pass filter with a roll-off of 40 db/decade

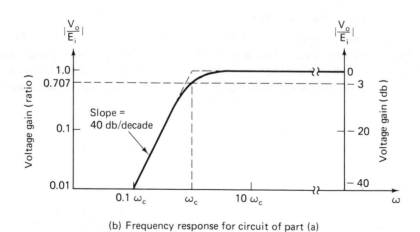

(b) Frequency response for circuit of part (a)

Figure 12-8 Circuit and frequency response for a 40-dB/decade high-pass Butterworth filter.

1. Choose a cutoff frequency, ω_c or f_c.
2. Let $C_1 = C_2 = C$ and choose a convenient value.
3. Calculate R_1 from

$$R_1 = \frac{1.414}{\omega_c C} \qquad (12\text{-}11)$$

4. $R_2 = \frac{1}{2}R_1$.   (12-12)
5. To minimize dc offset, let $R_f = R_1$.

Example 12-8

In Fig. 12-8(a), let $C_1 = C_2 = 0.01~\mu\text{F}$. Calculate (a) R_1 and (b) R_2 for a cutoff frequency of 1 kHz.
Solution. (a) From Eq. (12-11),

$$R_1 = \frac{1.414}{(6.28)(1 \times 10^3)(0.01 \times 10^{-6})} = 22.5~\text{k}\Omega$$

(b) $R_2 = \frac{1}{2}(22.5~\text{k}\Omega) = 11.3~\text{k}\Omega$.

Example 12-9

Calculate (a) R_1 and (b) R_2 in Fig. 12-8(a) for a cutoff frequency of 80 krad/s. $C_1 = C_2 = 125$ pF.
Solution. (a) From (12-11),

$$R_1 = \frac{1.414}{(80 \times 10^3)(125 \times 10^{-12})} = 140~\text{k}\Omega$$

(b) $R_2 = \frac{1}{2}(140~\text{k}\Omega) = 70~\text{k}\Omega$.

12-5.4 60-dB/Decade Filter

As with the low-pass filter of Fig. 12-5, a high-pass filter of 60 dB/decade can be constructed by cascading a 40-dB/decade filter with a 20-dB/decade filter. This circuit (like the other high- and low-pass filters) is designed as a Butterworth filter to have the frequency response in Fig. 12-9(b). The design steps for Fig. 12-9(a) are

1. Choose the cutoff frequency, ω_c or f_c.
2. Let $C_1 = C_2 = C_3 = C$ and choose a convenient value.
3. Calculate R_3 from

$$R_3 = \frac{1}{\omega_c C} \qquad (12\text{-}13)$$

4. Let $R_1 = 2R_3$.   (12-14)
5. Let $R_2 = \frac{1}{2}R_3$.   (12-15)
6. To minimize dc offset current, let $R_{f1} = R_1$ and $R_{f2} = R_3$.

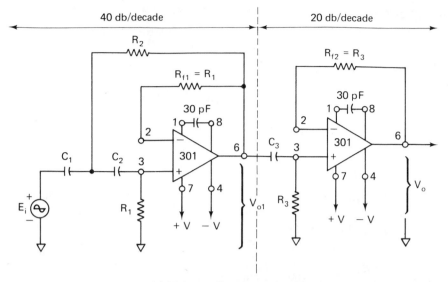

(a) High-pass filter for a 60-db/decade slope

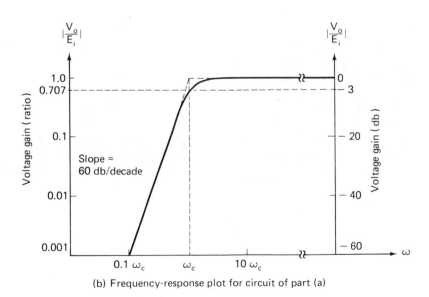

(b) Frequency-response plot for circuit of part (a)

Figure 12-9 Circuit and frequency response for a 60-dB/decade Butterworth high-pass filter.

Example 12-10

For Fig. 12-9(a), let $C_1 = C_2 = C_3 = C = 0.1 \; \mu\text{F}$. Determine (a) R_3, (b) R_1, and (c) R_2 for a cutoff frequency of 1 krad/s.
Solution. (a) By Eq. (12-13),

$$R_3 = \frac{1}{(1 \times 10^3)(0.1 \times 10^{-6})} = 10 \; \text{k}\Omega$$

(b) $R_1 = 2R_3 = 2(10 \; \text{k}\Omega) = 20 \; \text{k}\Omega$.
(c) $R_2 = \frac{1}{2}R_3 = \frac{1}{2}(10 \; \text{k}\Omega) = 5 \; \text{k}\Omega$.

Example 12-11

Determine (a) R_3, (b) R_1 and (c) R_2 in Fig. 12-9(a) for a cutoff frequency of 60 kHz. Let $C_1 = C_2 = C_3 = C = 220 \; \text{pF}$.
Solution. (a) From Eq. (12-13),

$$R_3 = \frac{1}{(6.28)(60 \times 10^3)(220 \times 10^{-12})} = 12 \; \text{k}\Omega$$

(b) $R_1 = 2R_3 = 2(12 \; \text{k}\Omega) = 24 \; \text{k}\Omega$.
(c) $R_2 = \frac{1}{2}R_3 = \frac{1}{2}(12 \; \text{k}\Omega) = 6 \; \text{k}\Omega$.

If desired, the 20-dB/decade section can come before the 40-dB/decade section, because the op amps provide isolation and do not load one another.

12-5.5   Comparison of Magnitudes and Phase Angles

Table 12-5 compares the magnitudes of the closed-loop gain for the three high-pass filters. For each increase in 20 dB/decade, the circuit not only has a steeper roll-off below ω_c but also remains closer to 0 dB above ω_c.

Table 12-5   Comparison of $|A_{CL}|$ for Figs. 12-7(a), 12-8(a), and 12-9(a)

ω	20 dB/decade; Fig. 12-7(a)	40 dB/decade; Fig. 12-8(a)	60 dB/decade; Fig. 12-9(a)
$0.1\omega_c$	0.1	0.01	0.001
$0.25\omega_c$	0.25	0.053	0.022
$0.5\omega_c$	0.445	0.24	0.124
ω_c	0.707	0.707	0.707
$2\omega_c$	0.89	0.97	0.992
$4\omega_c$	0.97	0.998	0.999
$10\omega_c$	1.0	1.0	1.0

The phase angle for a 20-dB/decade Butterworth high-pass filter is 45° at ω_c. For a 40-dB/decade filter it is 90°, and for a 60-dB/decade filter it is 135°.

Other phase angles in the vicinity of ω_c for the three filters are given in Table 12-6.

Table 12-6 Comparison of phase angles for Figs. 12-7(a), 12-8(a), and 12-9(a)

ω	20 dB/decade; Fig. 12-7(a)	40 dB/decade; Fig. 12-8(a)	60 dB/decade; Fig. 12-9(a)
$0.1\omega_c$	84°	172°	256°
$0.25\omega_c$	76°	143°	226°
$0.5\omega_c$	63°	137°	210°
ω_c	45°	90°	135°
$2\omega_c$	27°	43°	60°
$4\omega_c$	14°	21°	29°
$10\omega_c$	6°	8°	12°

12-6 BAND-PASS FILTERS

12-6.1 Introduction

A *band-pass* filter is a circuit designed to pass signals only in a certain band of frequencies while rejecting all signals outside this band. Figures 12-1(c) and 12-10(a) show the frequency response of a band-pass filter. This type of filter has a maximum output voltage V_{max}, or maximum voltage gain A_r, at one frequency called the *resonant frequency* ω_r. If the frequency varies from resonance, the output voltage decreases. There is one frequency above ω_r and one below ω_r at which the voltage gain is $0.707A_r$. These frequencies are designated by ω_h, the *high cutoff frequency*, and ω_l, the *low cutoff frequency*. The band of frequencies between ω_h and ω_l is the *bandwidth*, B:

$$B = \omega_h - \omega_l \tag{12-16}$$

Band-pass filters are classified as either narrow-band or wide-band. A narrow-band filter is one that has a bandwidth of less than one-tenth the resonant frequency ($B < 0.1\omega_r$). If the bandwidth is greater than one-tenth the resonant frequency ($B > 0.1\omega_r$), the filter is a wide-band filter. The ratio of resonant frequency to bandwidth is known as the *quality factor*, Q, of the circuit. Q indicates the selectivity of the circuit. The higher the value of Q, the more selective the circuit. In equation form,

$$Q = \frac{\omega_r}{B} \tag{12-17a}$$

or

$$B = \frac{\omega_r}{Q} \text{ rad/s} \tag{12-17b}$$

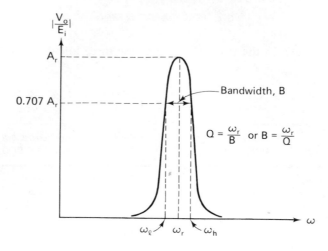

(a) Frequency response of a bandpass filter

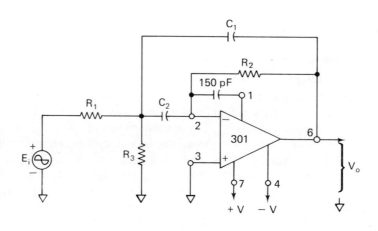

(b) Bandpass filter

Figure 12-10   Band-pass filter and frequency response.

For narrow-band filters, the Q of the circuit is greater than 10, and for wide-band filters, Q is less than 10.

12-6.2   Narrow-Band Band-Pass Filters

The circuit of Fig. 12-10(b) can be designed as either a wide-band filter ($Q < 10$) or as a narrow-band filter ($Q > 10$). Unlike either the low-pass or the high-pass filters of Sections 12-1 to 12-5, the filter of Fig. 12-10(b) can be designed for a closed-loop gain greater than 1. The maximum gain A_r occurs at the resonant frequency, as shown in Fig. 12-10(a). Normally, the designer of a band-pass filter first chooses the resonant frequency ω_r and the bandwidth B and calculates Q from Eq. (12-17a). For some designs, ω_r and Q are chosen and the bandwidth B is calculated from Eq. (12-17b). To simplify the design and reduce the number of calculations, choose $C_1 = C_2 = C$ and solve for R_1, R_2, and R_3 from the following equations:

$$R_2 = \frac{2}{BC} \tag{12-18}$$

$$R_1 = \frac{R_2}{2A_r} \tag{12-19}$$

$$R_3 = \frac{R_2}{4Q^2 - 2A_r} \tag{12-20}$$

To guarantee that R_3 is a positive value, be sure that $4Q^2 > 2A_r$. B in Eq. (12-18) is in radians per second.

Example 12-12

Design the band-pass filter of Fig. 12-10(b) to have $\omega_r = 10$ krad/s, $A_r = 40$, $Q = 20$, and $C_1 = C_2 = C = 0.01 \ \mu F$.

Solution. By Eq. (12-17b),

$$B = \frac{10 \times 10^3}{20} = 0.5 \text{ krad/s}$$

From Eqs. (12-18) to (12-20),

$$R_2 = \frac{2}{(0.5 \times 10^3)(0.01 \times 10^{-6})} = 400 \text{ k}\Omega$$

$$R_1 = \frac{400 \times 10^3}{2(40)} = 5 \text{ k}\Omega$$

$$R_3 = \frac{400 \times 10^3}{4(400) - 2(40)} = 263 \ \Omega$$

Example 12-13

If the bandwidth of Example 12-12 is to be increased to 1 krad/sec, calculate (a) Q; (b) R_2; (c) R_1; (d) R_3.

Solution. (a) By Eq. (12-17a),

$$Q = \frac{10 \times 10^3}{1 \times 10^3} = 10$$

(b) By Eq. (12-18),

$$R_2 = \frac{2}{(1 \times 10^3)(0.01 \times 10^{-6})} = 200 \text{ k}\Omega$$

(c) By Eq. (12-19),

$$R_1 = \frac{200 \times 10^3}{2(40)} = 2.5 \text{ k}\Omega$$

(d) By Eq. (12-20),

$$R_3 = \frac{200 \times 10^3}{4(100) - 2(40)} = 625\ \Omega$$

12-6.3 Wide-Band Filters

As stated previously, a wide-band band-pass filter is a circuit in which $Q < 10$. The circuit of Fig. 12-10(b) can be designed as a wide-band filter, and Eqs. (12-18) to (12-20) can be used, provided that $4Q^2 > 2A_r$.

Example 12-14

Design Fig. 12-10(b) to have $\omega_r = 20$ krad/s, $A_r = 10$, $Q = 5$, and $C_1 = C_2 = C = 0.01\ \mu$F.

Solution. By Eq. (12-17b),

$$B = \frac{20 \times 10^3}{5} = 4\ \text{krad/s}$$

From Eqs. (12-18) to (12-20),

$$R_2 = \frac{2}{(4 \times 10^3)(0.01 \times 10^{-6})} = 50\ \text{k}\Omega$$

$$R_1 = \frac{50 \times 10^3}{2(10)} = 2.5\ \text{k}\Omega$$

$$R_3 = \frac{50 \times 10^3}{4(25) - 2(10)} = 625\ \Omega$$

Another idea for a wide-band filter is to connect a low-pass filter to a high-pass filter. For example, the low-pass filter of Example 12-5 connected to the high-pass filter of Example 12-10 gives the frequency response shown in Fig. 12-11. Although this wide-band filter uses four op amps, the roll-off is 60 dB/

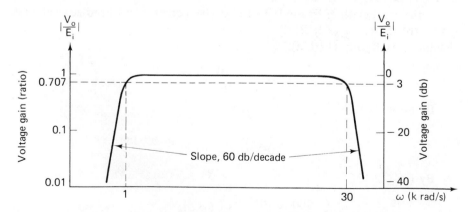

Figure 12-11 Wide-band filter response obtained by connecting Fig. 12-5(a) to Fig. 12-9(a).

decade at both the high and low cutoff frequencies. The gain in the pass band is 1, because the gain for both the low- and high-pass filter is 1. When one is building this type of wide-band filter, it makes no difference which filter comes first.

12-7   NOTCH FILTERS

The circuit of Fig. 12-12(a) is a *notch* or *band-elimination filter*. Its frequency-response curve is shown in Figs. 12-1(d) and 12-12(b). Undesired frequencies are attenuated in the stop band. For example, it may be necessary to attenuate 60-Hz or 400-Hz noise signals induced in a circuit by motor generators. Design

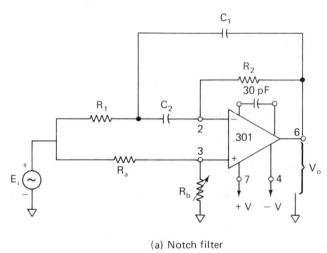

(a) Notch filter

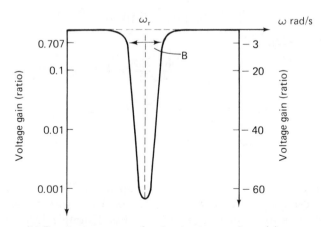

(b) Frequency response for the circuit part of part (a)

Figure 12-12   Circuit and frequency response for a notch filter.

of the notch filter is carried out in five steps. You usually know or are designing for a required bandwidth B or Q and resonant frequency ω_c. Then proceed as follows:

1. Choose $C_1 = C_2 = C$ (some convenient value).
2. Calculate R_2 from

$$R_2 = \frac{2}{BC} \tag{12-21}$$

   where B is in radians per second.
3. Calculate R_1 from

$$R_1 = \frac{R_2}{4Q^2} \tag{12-22}$$

4. Choose for R_a a convenient value, such as 1 kΩ.
5. Calculate R_b from

$$R_b = 2Q^2 R_a \tag{12-23}$$

This procedure is illustrated by an example.

Example 12-15

Design a notch filter from Fig. 12-12(a) for $f_r = 400$ Hz and $Q = 5$. Let $C_1 = C_2 = C = 0.01 \; \mu$F.
Solution. $\omega_r = 2\pi f_r = (6.28)(400) = 2.51$ krad/s. From Eq. (12-17a),

$$B = \frac{2.51 \times 10^3}{5} \cong 500 \text{ rad/s}$$

From Eq. (12-21),

$$R_2 = \frac{2}{(500)(0.01 \times 10^{-6})} = 400 \text{ k}\Omega$$

From Eq. (12-22),

$$R_1 = \frac{400 \text{ k}\Omega}{4(25)} = 4 \text{ k}\Omega$$

Choose $R_a = 1$ kΩ and from Eq. (12-23); $R_b = 2(25)1 \text{ k}\Omega = 50 \text{ k}\Omega$.

When building the notch filter of Fig. 12-12 the following procedure should be used:

1. Ground the $(+)$ terminal of the op amp. The resulting network is a band-pass filter similar to Fig. 12-10(b) but without R_3. The gain for this band-pass filter at ω_r is $2Q^2$. (For Example 12-15, the gain is 50.) Adjust R_1 and R_2 to fine-tune ω_r and B.

2. Remove the ground at the (+) input and adjust R_b to the value obtained from Eq. (12-23).

The frequency response of a notch filter is shown in Fig. 12-12(b). Note that the bandwidth is still that band of frequencies at -3 dB from the maximum value.

PROBLEMS

12-1. List the four types of filters.

12-2. What type of filter has a constant output voltage from dc up to the cutoff frequency?

12-3. What is a filter called that passes a band of frequencies while attenuating all frequencies outside the band?

12-4. In Fig. 12-2(a), if $R = 100$ kΩ and $C = 0.02$ μF, what is the cutoff frequency?

12-5. The low-pass filter of Fig. 12-2(a) is to be designed for a cutoff frequency of 4.5 kHz. If $C = 0.005$ μF, calculate R.

12-6. If the cutoff frequency in Fig. 12-2(a) is 50 krad/s and $R = 20$ kΩ, determine C.

12-7. What are the two characteristics of a Butterworth filter?

12-8. Design a -40-dB/decade low-pass filter at a cutoff frequency of 10 krad/s. Let $R_1 = R_2 = 50$ kΩ.

12-9. In Fig. 12-4(a), if $R_1 = R_2 = 10$ kΩ, $C_1 = 0.001$ μF, and $C_2 = 0.002$ μF, calculate the cutoff frequency f_c.

12-10. Calculate (a) C_3, (b) C_1, and (c) C_2 in Fig. 12-5(a) for a cutoff frequency of 10 krad/s. $R_1 = R_2 = R_3 = 10$ kΩ.

12-11. If $R_1 = R_2 = R_3 = 20$ kΩ, $C_1 = 0.002$ μF, $C_2 = 0.008$ μF, and $C_3 = 0.004$ μF in Fig. 12-5(a), determine the cutoff frequency ω_c.

12-12. In Fig. 12-5(a), $C_1 = 0.01$ μF, $C_2 = 0.04$ μF, and $C_3 = 0.02$ μF. Calculate R for a cutoff frequency of 1 kHz.

12-13. Calculate R in Fig. 12-7(a) if $C = 0.04$ μF and $f_c = 500$ Hz.

12-14. In Fig. 12-7(a) calculate (a) ω_c and (b) f_c if $R = 10$ kΩ and $C = 0.001$ μF.

12-15. Design a 40-dB/decade high-pass filter for $\omega_c = 5$ krad/s. $C_1 = C_2 = 0.02$ μF.

12-16. Calculate (a) R_1 and (b) R_2 in Fig. 12-8(a) for a cutoff frequency of 40 krad/s. $C_1 = C_2 = 250$ pF.

12-17. For Fig. 12-9(a), let $C_1 = C_2 = C_3 = 0.05$ μF. Determine (a) R_3, (b) R_1, and (c) R_2 for a cutoff frequency of 500 Hz.

12-18. The circuit of Fig. 12-9(a) is designed with the values $C_1 = C_2 = C_3 = 400$ pF, $R_1 = 100$ kΩ, $R_2 = 25$ kΩ, and $R_3 = 50$ kΩ. Calculate the cutoff frequency f_c.

12-19. If $\omega_h = 22.5$ krad/s and $\omega_l = 22.1$ krad/s, what is the bandwidth (a) in rad/s; (b) in hertz?

12-20. For the values given in Problem 12-19, determine the quality factor Q.

12-21. Design the band-pass filter of Fig. 12-10(b) to have $\omega_r = 10$ krad/s, $A_r = 5$, $Q = 10$, and $C_1 = C_2 = 0.001$ μF.

12-22. If the gain at the resonant frequency in Problem 12-21 is increased to 10, what are the values for R_1, R_2, and R_3?

12-23. In the band-pass filter of Fig. 12-10(b), $C_1 = C_2 = 0.01$ μF, $R_1 = 40$ kΩ, $R_2 = 400$ kΩ, and $R_3 = 252$ Ω; determine (a) bandwidth (rad/s); (b) the gain at the resonant frequency; (c) Q; (d) the resonant frequency in hertz.

12-24. Design the notch filter of Fig. 12-12(a) to have $\omega_r = 2$ krad/s, $Q = 10$, and $C_1 = C_2 = 0.1$ μF. Let $R_a = 1$ kΩ.

integrated-
circuit timers

13

13-0 INTRODUCTION

Applications such as oscillators, pulse generators, ramp or square-wave generators, one-shot multivibrators, burglar alarms, and voltage monitors all require
a circuit capable of producing timing intervals. The most popular integrated-
circuit timer is the 555, first introduced by Signetics Corporation (see Appendix
4). Similar to general-purpose op amps, the 555 is reliable, easy to use in a
variety of applications, and low in cost. The 555 can also operate from supply
voltages of 5 V to +18 V, making it compatible with both TTL (transistor-
transistor logic) circuits and op amp circuits. The 555 timer can be considered
a functional block that contains two comparators, two transistors, three equal
resistors, a flip-flop, and an output stage. These are shown in Fig. 13-1.

Besides the 555 timer, there are also available counter timers such as Exar's
XR—2240 (see Appendix 5). The 2240 contains a 555 timer plus a programmable
binary counter in a single 16-pin package. A single 555 has a maximum timing
range of approximately 15 min. Counter timers have a maximum timing range
of days. The timing range of both can be extended to months or even years by
cascading. Our study of timers will begin with the 555 and its applications and
then proceed to the counter timers.

13-1 OPERATING MODES OF THE 555 TIMER

The 555 IC timer has two modes of operation, either as an astable (free-running)
multivibrator or as a monostable (one-shot) multivibrator. Free-running operation of the 555 is shown in Fig. 13-2(a). The output voltage switches from a

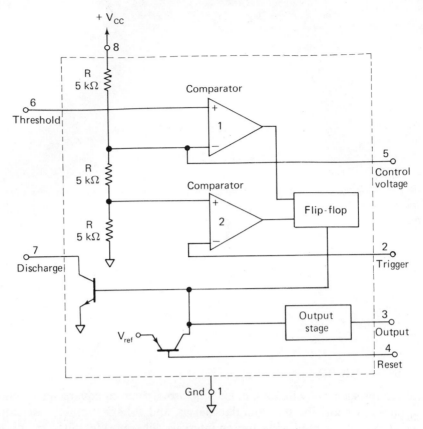

Figure 13-1 A 555 integrated-circuit timer.

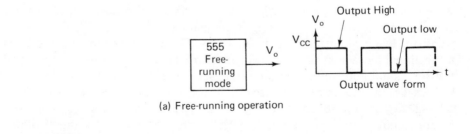

(a) Free-running operation

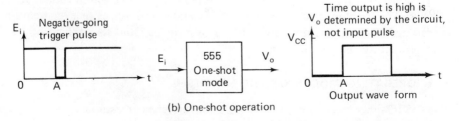

(b) One-shot operation

Figure 13-2 Operating modes of a 555 timer.

274

high to a low state and back again. The time the output is either high or low is determined by a resistor–capacitor network connected externally to the 555 timer (see Section 13-2). The value of the high output voltage is slightly less than V_{CC}. The value of the output voltage in the low state is approximately 0.1 V.

When the timer is operated as a one-shot multivibrator, the output voltage is low until a negative-going trigger pulse is applied to the timer; then the output switches high. The time the output is high is determined by a resistor and capacitor connected to the IC timer. At the end of the timing interval, the output returns to the low state. Monostable operation is examined further in Sections 13-5 and 13-6.

To understand how a 555 timer operates, a brief description of each terminal is given in Section 13-2.

13-2   TERMINALS OF THE 555

13-2.1   Packaging and Power Supply Terminals

The 555 timer is available in two package styles, TO 99 and DIP, as shown in Fig. 13-3a and Appendix 4. Pin 1 is the common, or ground, terminal, and pin 8 is the positive voltage supply terminal V_{CC}. V_{CC} can be any voltage between +5 V and +18 V. Thus the 555 can be powered by existing digital logic supplies (+5 V), linear IC supplies (+15 V), and automobile or dry cell batteries. Internal circuitry requires about 0.7 mA per supply volt (10 mA for $V_{CC} = +15$) to set up internal bias currents. Maximum power dissipation for the package is 600 mW.

13-2.2   Output Terminal

As shown in Fig. 13-3(b) and (c), the output terminal, pin 3, can either source or sink current. A *floating* supply load is *on* when the *output* is *low*, and *off* when the output is *high*. A *grounded* load is *on* when the *output* is *high*, and *off* when the output is *low*. In normal operation either a supply load or a grounded load is connected to pin 3. Most applications do not require both types of loads at the same time.

The maximum sink or source current is technically 200 mA, but more realistically is 40 mA. The high output voltage [Fig. 13-3(c)] is about 0.5 V below V_{CC}, and the low output voltage [Fig. 13-3(b)] is about 0.1 V above ground, for load currents below 25 mA.

13-2.3   Reset Terminal

The reset terminal, pin 4, allows the 555 to be disabled and override command signals on the trigger input. When not used, the reset terminal should be wired to $+V_{CC}$. If the reset terminal is grounded or its potential reduced below 0.4 V, both the output terminal, pin 3, and the discharge terminal, pin 7, are

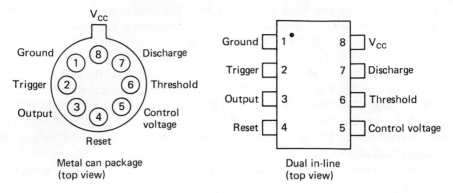

(a) 555 pin connections and package styles

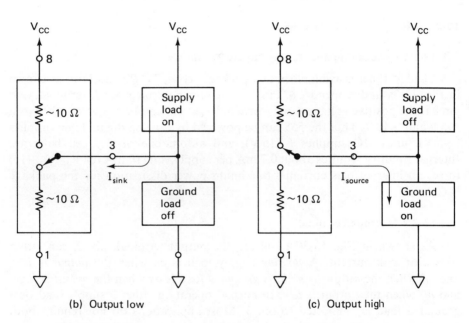

Figure 13-3 The 555 timer-output operation and package terminals. Either a grounded or a supply load can be connected, although usually not simultaneously.

at approximately ground potential. In other words, the output is held low. If the output was high, a ground on the reset terminal immediately forces the output low.

13-2.4   Discharge Terminal

Discharge terminal, pin 7, is used to discharge an external timing capacitor during the time the output is low. When the output is high, pin 7 acts as an open circuit and allows the capacitor to charge at a rate determined by an external resistor or resistors and capacitor. Figure 13-4 shows a model of the discharge terminal for when C is discharging and for when C is charging.

13-2.5 Control Voltage Terminal

A 0.01-μF filter capacitor is usually connected from the control voltage terminal, pin 5, to ground. The capacitor bypasses noise and/or ripple voltages from the power supply to minimize their effect on threshold voltage. The control voltage terminal may also be used to change both the threshold and trigger voltage levels. For example, connecting a 5-kΩ resistor between pins 5 and 8 changes threshold voltage to $0.8V_{CC}$ and the trigger voltage to $0.4V_{CC}$. An

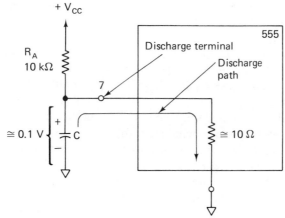

(a) Model of the discharge terminal when the output is low, and capacitor is discharging

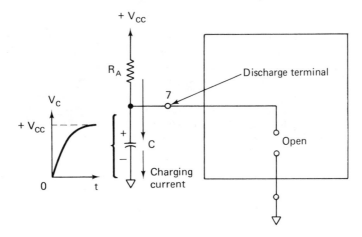

(b) Model of the discharge terminal when the output is high, and capacitor is charging

Figure 13-4 Operation of discharge terminal.

external voltage applied to pin 5 will change both threshold and trigger voltages and can also be used to modulate the output waveform.

13-2.6 Trigger and Threshold Terminals

The 555 has two possible operating states and two possible memory states. They are determined by *both* the *trigger* input, pin 2, and the *threshold* input, pin 6. The trigger input is compared by comparator 1 in Fig. 13-1, with a lower threshold voltage V_{LT} that is equal to $V_{CC}/3$. The threshold input is compared by comparator 1 with a higher threshold voltage V_{UT} that is equal to $2V_{CC}/3$. Each input has two possible voltage levels, either above or below its reference voltage. Thus with two inputs there are four possible combinations that will cause four possible operating states.

The four possible input combinations and corresponding states of the 555 are given in Table 13-1. In operating state A, *both* trigger and threshold are *below* their respective threshold voltages and the output terminal (pin 3) is *high*. In operating state D, *both* inputs are *above* their threshold voltages and the output terminal is *low*.

The observation that low inputs give a high output, and high inputs give a low output, might lead you to conclude that the 555 acts as an inverter. However, as shown in Table 13-1, the 555 also has two memory states. State B occurs when the trigger input is *below*, *and* the threshold input is *above* their respective reference voltages. Memory state C occurs when the trigger input is *above*, *and* the threshold input is *below* their respective reference voltages.

Table 13-1 Operating states of a 555 timer: $V_{UT} = 2V_{CC}/3$, $V_{LT} = V_{CC}/3$; high $\cong V_{CC}$, low or ground $\cong$ 0 V

Operating state	Trigger pin 2	Threshold pin 6	State of terminals	
			Output 3	Discharge 7
A	Below V_{LT}	Below V_{UT}	High	Open
B	Below V_{LT}	Above V_{UT}	Remembers last state	
C	Above V_{LT}	Below V_{UT}	Remembers last state	
D	Above V_{LT}	Above V_{UT}	Low	Ground

A visual aid in understanding how these operating states occur is presented in Fig. 13-5. An input voltage E_i is applied to *both* trigger and threshold input terminals. When E_i is below V_{LT} during time intervals A–B and E–F, state A operation results, so that output V_{03} is high. When E_i lies above V_{LT} but below V_{UT}, within time B–C, the 555 enters state C and remembers its last A state. When E_i exceeds V_{UT}, state D operation sends the output low. When E_i drops between V_{UT} and V_{LT} during time D–E, the 555 remembers the last D state and its output stays low. Finally, when E_i drops below V_{LT} during time E–F, the A state sends the output high.

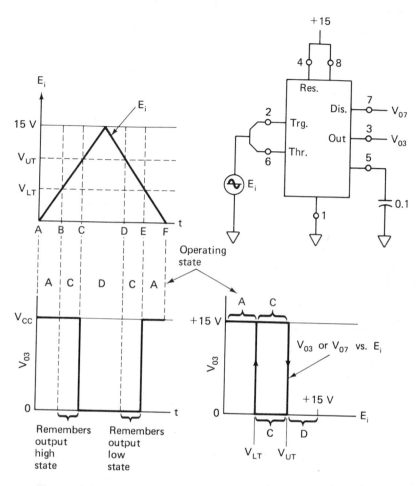

Figure 13-5 Three of the four operating states of a 555 timer are shown by a test circuit to measure E_i and V_{03} versus time and V_{03} versus E_i.

By plotting output V_{03} against E_i in Fig. 13-5, we see a hysteresis characteristic. Recall from Chapter 4 that a hysteresis loop means that the circuit has memory. This also means that if the inputs are in one of the memory states, you cannot tell what state the output is now in, unless you know the previous state. Two *power-on* applications will now be given to show how to analyze circuit operation from Table 13-1.

13-2.7 Power-on Time Delays

There are two types of timing events that may be required during a power-on application. You may wish to apply power to one part of a system and wait for a short interval before starting some other part of a system. A circuit that solves this problem is shown in Fig. 13-6(a). When the power switch is thrown to on

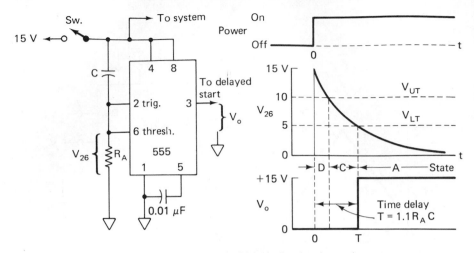

(a) Output V_o does not go high until a time interval
T elapses after application of power at t = 0

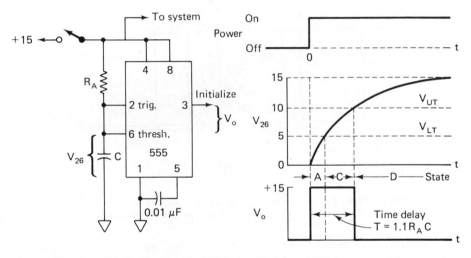

(b) Output V_o goes high for a time interval T
after power is applied

Figure 13-6 Power-on time-delay applications are analyzed by reference to
Table 13-1.

at $t = 0$, the initial capacitor voltage is zero. Therefore, both pins 2 and 6 are
above their respective thresholds and the output stays low in operating state D.
As capacitor C charges, threshold drops below V_{UT} while trigger is still above
V_{LT}, forcing the 555 into memory state C. Finally, both trigger and threshold
drop just below V_{LT}, where the 555 enters state A and forces the output high at
time T.

The net result is that an output from pin 3 of the 555 is delayed for a time interval T after the switch closure at $t = 0$. The time delay is found from $T = 1.1R_AC$.

By interchanging R_A and C, a time delay with a high output can be generated. In the circuit of Fig. 13-6(b), power is applied to a system when the switch is closed. The 555's output goes high for a period of time T and then goes low. T is found from Eq. (13-9). This type of startup pulse is typically used to reset counters and initialize computer sequences after a power failure. It also can allow time for an operator to exit after an alarm system has been turned on before arming the system.

13-3 FREE-RUNNING OR ASTABLE OPERATION

13-3.1 Circuit Operation

The 555 is connected as a free-running multivibrator in Fig. 13-7(a). Refer to the waveshapes in Fig. 13-7(b) to follow the circuit's operation. At time A both pins 2 and 6 go just below $V_{LT} = \frac{1}{3}V_{CC}$ and output pin 3 goes high (state A). Pin 7 also becomes an open, so capacitor C charges through $R_A + R_B$. During time A–B, the 555 is in memory state C, remembering the previous state. When V_C goes just above $V_{UT} = \frac{2}{3}V_{CC}$ at time B, the 555 enters state D and sends the output low. Pin 7 also goes low and capacitor C discharges through resistor R_B. During time B–C the 555 is in memory state C, remembering the previous state D. When V_C drops just below V_{LT}, the sequence repeats.

13-3.2 Frequency of Oscillation

The output stays high during the time interval that C charges from $\frac{1}{3}V_{CC}$ to $\frac{2}{3}V_{CC}$ as shown in Fig. 13-7(b) and (c). This time interval is given by

$$t_{high} = 0.695(R_A + R_B)C \qquad (13\text{-}1)$$

The output is low during the time interval that C discharges from $\frac{2}{3}V_{CC}$ to $\frac{1}{3}V_{CC}$ and is given by

$$t_{low} = 0.695R_BC \qquad (13\text{-}2)$$

Thus the total period of oscillation T is

$$T = t_{high} + t_{low} = 0.695(R_A + 2R_B)C \qquad (13\text{-}3)$$

The free-running frequency of oscillation f is

$$f = \frac{1}{T} = \frac{1.44}{(R_A + 2R_B)C} \qquad (13\text{-}4)$$

Figure 13-7(c) is a plot of Eq. (13-4) for different values of $(R_A + 2R_B)$ and

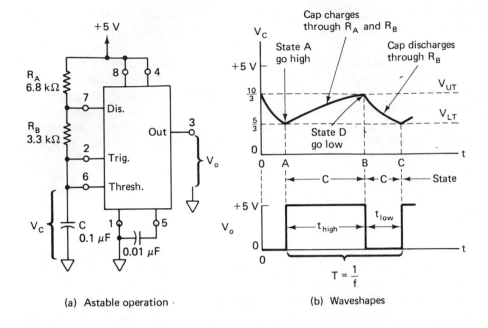

(a) Astable operation

(b) Waveshapes

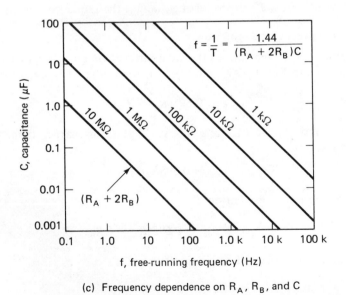

(c) Frequency dependence on R_A, R_B, and C

Figure 13-7 Waveshapes are shown in (b) for the free-running astable multivibrator in (a). Frequency of operation is determined by the resistance and capacitance values in (c).

quickly shows what combinations of resistance and capacitance are needed to design an astable multivibrator.

Example 13-1

Calculate (a) t_{high}, (b) t_{low}, and (c) the free-running frequency for the timer circuit of Fig. 13-7(a).

Solution. (a) By Eq. (13-1),

$$t_{high} = 0.695(6.8 \text{ k}\Omega + 3.3 \text{ k}\Omega)(0.1 \ \mu\text{F}) = 0.7 \text{ ms}$$

(b) By Eq. (13-2),

$$t_{low} = 0.695(3.3 \text{ k}\Omega)(0.1 \ \mu\text{F}) = 0.23 \text{ ms}$$

(c) By Eq. (13-4),

$$f = \frac{1.44}{(6.8 \text{ k}\Omega) + (2)(3.3 \text{ k}\Omega)(0.1 \ \mu\text{F})} = 1.07 \text{ kHz}$$

The answer to part (c) agrees with results obtainable from Fig. 13-7(c).

13-3.3 Duty Cycle

The ratio of time when the output is low t_{low} to the total period T is called the *duty cycle D*. In equation form,

$$D = \frac{t_{low}}{T} = \frac{R_B}{R_A + 2R_B} \tag{13-5}$$

Example 13-2

Calculate the duty cycle for the values given in Fig. 13-7(a).

Solution. By Eq. (13-5),

$$D = \frac{3.3 \text{ k}\Omega}{6.8 \text{ k}\Omega + 2(3.3 \text{ k}\Omega)} = 0.25$$

This checks with Fig. 13-6(b) which shows that the timer's output is low for approximately 25% of the total period T. Equation (13-5) shows that it is impossible to obtain a duty cycle of $\frac{1}{2}$ or 50%. As presented, the circuit of Fig. 13-7(a) is not capable of producing a square wave. The only way D in Eq. (13-5) can equal $\frac{1}{2}$ is for R_A to equal 0. Then there would be a short between V_{CC} and pin 7. However, R_A must be large enough so that when the discharge transistor is "on," current through it is limited to 0.2 A. Thus the minimum value of R_A in ohms is given by

$$\text{minimum } R_A \cong \frac{V_{CC}}{0.2 \text{ A}} \tag{13-6a}$$

or

$$R_A \cong 5V_{CC} \tag{13-6b}$$

The conclusion drawn from Eq. (13-6b) is that R_A cannot be equal to 0. Therefore, to extend the duty cycle an alternative solution must be found.

13-3.4   Extending the Duty Cycle

The duty cycle for the circuit of Fig. 13-7(a) can never be equal to or greater than 50%, as discussed in Section 13-3.3. By connecting a diode in parallel with R_B in Fig. 13-8(a), a duty cycle of 50% or greater can be obtained. Now the capacitor charges through R_A and the diode, but discharges through R_B. The times for the output waveform are

$$t_{high} = 0.695 R_A C \tag{13-7a}$$

$$t_{low} = 0.695 R_B C \tag{13-7b}$$

$$T = 0.695(R_A + R_B)C \tag{13-7c}$$

Equations (13-7a) and (13-7b) show that if $R_A = R_B$, then the duty cycle is 50%, as shown in Fig. 13-8(b) and (c).

13-4   APPLICATIONS OF THE 555 AS AN ASTABLE MULTIVIBRATOR

13-4.1   Tone-Burst Oscillator

With the switch in Fig. 13-9 set to the "continuous" position, the B 555 timer functions as a free-running multivibrator. The frequency can be varied from about 1.3 kHz to 14 kHz by the 10-kΩ potentiometer. If the potentiometer is replaced by a thermistor or photoconductive cell, the oscillating frequency will be proportional to temperature or light intensity respectively.

The A 555 timer oscillates at a slower frequency. The 1-MΩ potentiometer sets the lowest frequency at about 1.5 Hz. Lower frequencies are possible by replacing the 1-μF capacitor with a larger value. When the connecting switch is thrown to the "burst" position, output pin 3 of the A timer alternately places a ground or high voltage on reset pin 4 of the B 555 timer. When pin 4 of the B timer is grounded, it cannnot oscillate, and when ungrounded the timer oscillates. This causes the B timer to oscillate in bursts. The output of the tone-burst generator is V_o and is taken from pin 3 of timer B. V_o can drive either an audio amplifier or a stepdown transformer directly to a speaker.

The 556 IC timer contains two 555 timers in a single 14-pin dual-in-line package. The tone-burst generator can be made with one 556.

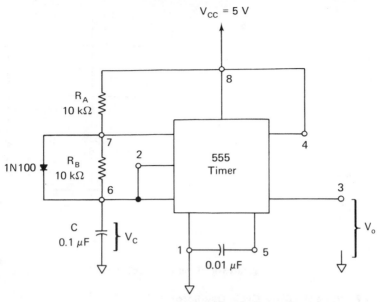

(a) Timer circuit to produce a 50% duty cycle

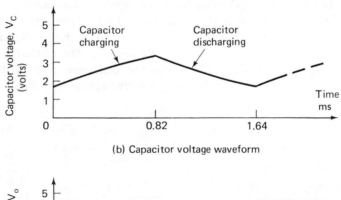

(b) Capacitor voltage waveform

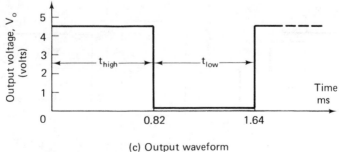

(c) Output waveform

Figure 13-8 Connecting a diode across R_B to produce duty cycles = 50%.

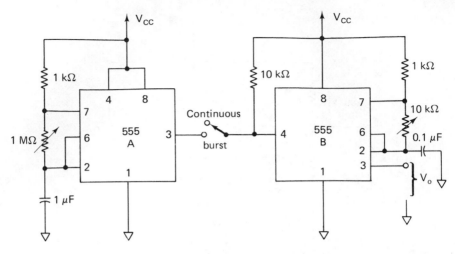

Figure 13-9   Tone-burst oscillator.

13-4.2   Variable-Duty-Cycle Oscillator

By adding another diode, resistor, and potentiometer to the 50% duty cycle circuit of Fig. 13-8(a), a variable duty-cycle square-wave generator can be constructed. The result is shown in Fig. 13-10, where independent charge and discharge paths for capacitor C are established by diodes D_A and D_B. The charge path for C is from V_{CC} through R_A and D_A. The discharge path for C is through D_B, R_B, and pin 7. The charge and discharge time t_{high} and t_{low} are given by Eqs. (13-7a) and (13-7b), respectively. The period T and duty cycle

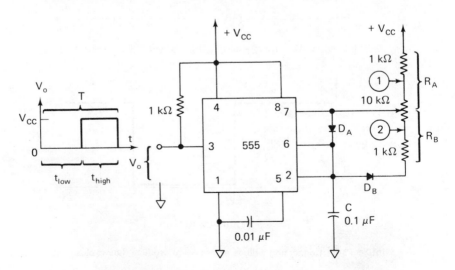

Figure 13-10   Variable-duty-cycle square-wave generator.

are given by

$$T = 0.7(R_A + R_B)C \qquad (13\text{-}8a)$$

$$\text{duty cycle} = \frac{R_B}{R_A + R_B} \qquad (13\text{-}8b)$$

Duty cycles from 1 to 99% can be realized.

Example 13-3

Calculate the duty cycle for Fig. 13-10 if the wiper of the 10-kΩ potentiometer is set at (a) position 1; (b) position 2.

Solution. (a) $R_A = 1$ kΩ and $R_B = 11$ kΩ, so

$$D = \frac{11 \text{ k}\Omega}{1 \text{ k}\Omega + 11 \text{ k}\Omega} \cong 0.92 \text{ or } 92\%$$

which means t_{low} is 92% of the total period T.
(b) $R_A = 11$ kΩ and $R_B = 1$ kΩ, so

$$D = \frac{1 \text{ k}\Omega}{11 \text{ k}\Omega + 1 \text{ k}\Omega} \cong 0.082 \text{ or } 8.3\%$$

In this position t_{low} is only 8.3% of the total period T.

13-5   ONE-SHOT OR MONOSTABLE OPERATION

13-5.1   Introduction

Not all applications require a continuous repetitive wave such as that obtained from a free-running multivibrator. Many applications need to operate only for a specified length of time. These circuits require a one-shot or monostable multivibrator. Figure 13-11(a) is a circuit diagram using the 555 for monostable operation. When a negative-going pulse is applied to pin 2, the output goes high and terminal 7 removes a short circuit from capacitor C. The voltage across C rises at a rate determined by R_A and C. When the capacitor voltage reaches $\frac{2}{3}V_{\text{CC}}$, comparator 1 in Fig. 13-1 causes the output to switch from high to low. The input and output voltage waveforms are shown in Fig. 13-11(a). The output is high for a time given by

$$t_{\text{high}} = 1.1R_A C \qquad (13\text{-}9)$$

Figure 13-11(b) is a plot of Eq. (13-9) and quickly shows the wide range of output pulses that are obtainable and the required values of R_A and C.

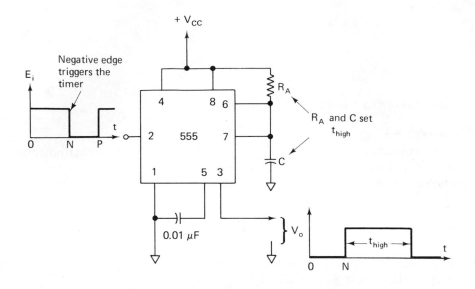

(a) 555 timer wired for monostable operation. Peak value
of E_i must be greater than or equal to $\frac{2}{3} V_{CC}$

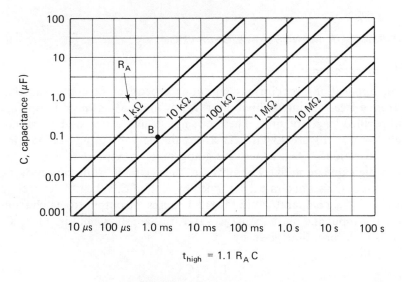

$$t_{high} = 1.1\, R_A\, C$$

(b) Design aid to determine output pulse duration

Figure 13-11   Monostable operation.

Example 13-4

If $R_A = 9.1$ kΩ, find C for an output pulse duration of 1 ms.
Solution. Rearrange Eq. (13-9):

$$C = \frac{t_{high}}{1.1R_A} = \frac{1 \times 10^{-3}s}{1.1(9.1 \times 10^3)\Omega} = 0.1 \ \mu F$$

This answer checks with that obtainable at point B in Fig. 13-11(b). For the 555 timer to trigger properly in this type of operation, the width of the trigger pulse must be less than t_{high} and a trigger input pulse network is needed so that the output does not switch on the positive-going edge of the trigger pulse (point P).

13-5.2 Input Pulse Circuit

Figure 13-12 shows the multivibrator wired for monostable operation [as in Fig. 13-11(a)]. R_i, C_i, and diode D are needed to generate a single output pulse for one input pulse.

Resistor R_A and capacitor C determine the time that the output is high, as given by Eq. (13-9). Resistor R_i is connected between V_{CC} and pin 2 to ensure that the output is low. C_i is charged to $(V_{CC} - E_i)$ until the negative trigger pulse occurs. The time constant of R_i and C_i should be small with respect to the output timing interval t_{high}. Diode D prevents the 555 timer from triggering on the positive-going edges of E_i. Waveforms for the input pulse, E_i, the pulse at pin 2, V_2, and the output pulse, V_0, are all shown in Fig. 13-12.

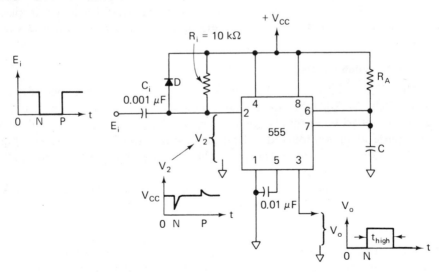

Figure 13-12 For satisfactory monostable operation, the input pulse network of R_i, C_i, and D is needed.

Example 13-5

(a) If $R_A = 10$ kΩ and $C = 0.2$ μF in Fig. 13-12, find t_{high}. (b) What is the time constant of R_i and C_i in Fig. 13-12?

Solution. (a) By Eq. (13-9),

$$t_{high} = 1.1(10 \times 10^3)(0.2 \times 10^{-6}) = 2.2 \text{ ms}$$

(b) Time constant $= R_i C_i = (10 \times 10^3)(0.001 \times 10^{-6}) = 0.01$ ms.

As with astable operation, the reset terminal pin 4 is normally tied to the supply voltage, V_{CC}. If pin 4 is grounded at any time, the timing cycle is stopped. When the reset terminal is grounded, both output pin 3 and discharge terminal 7 go to ground potential. Thus the output goes low and any charge accumulated by the timing capacitor C is removed. As long as the reset terminal is grounded, these conditions remain.

13-6 APPLICATIONS OF THE 555 AS A ONE-SHOT MULTIVIBRATOR

13-6.1 Water-Level Fill Control

In Fig. 13-13(a), the start switch is closed and the output of the 555 is low. When the start switch is opened, the output goes high to actuate the pump. The time interval the output is high is given by Eq. (13-9). Upon completion of the timing interval, the output of the 555 returns to its low state, turning the pump off. The height of the water level is set by the timing interval which is set by R_A and C. In the event of a potential overflow, the overfill switch must place a ground on reset pin 4, which causes the timer's output to go low and stops the pump.

13-6.2 Touch Switch

The 555 is wired as a one-shot multivibrator in Fig. 13-13(b) to perform as a touch switch. A 22-MΩ resistor to pin 2 holds the 555 in its idle state. If you scuff your feet to build up a static charge, the 555 will produce a single shot output pulse when you touch the finger plate. If the electrical noise level is high (due, for example, to fluorescent lights) the 555 may oscillate when you touch the finger plate. Reliable and consistent triggering will occur if a thumb is placed on a ground plate and fingers of the same hand tap the finger plate. An isolated power supply or batteries should be used for safety.

13-6.3 Frequency Divider

Figure 13-11(a) can be used as a frequency divider if the timing interval is adjusted to be longer than the period of the input signal E_i. For example, suppose that the frequency of E_i is 1 kHz, so that its period is 1 ms. If $R_A = 10$ kΩ and $C = 0.1$ μF, the timing interval given by Eq. (13-9) is $t_{high} = 1.1$ ms. Therefore, the one shot will be triggered by the first negative-going pulse of

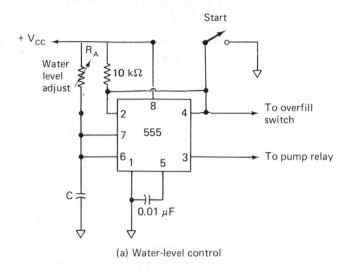

(a) Water-level control

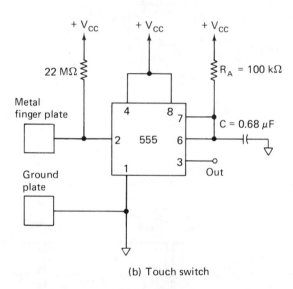

(b) Touch switch

Figure 13-13 Basic one-shot applications of the 555.

E_i, but the output will still be high when the second negative-going pulse occurs. The one-shot will, however, be retriggered on the third negative-going pulse. In this example, the one-shot triggers on every other pulse of E_i, so there is only one output for every two input pulses; thus E_i is divided by 2.

Example 13-6

(a) Calculate the timing interval in Fig. 13-11(a) if $R_A = 10\ \text{k}\Omega$ and $C = 0.1\ \mu\text{F}$. (b) What value of R_A should be installed to divide a 1-kHz input signal by 3?

Solution. (a) By Eq. (13-9), $t_{high} = 1.1(10 \times 10^3)(0.1 \times 10^{-6}) = 1.1$ ms. (b) t_{high} should exceed two periods of E_i, or 2 ms, and be less than three periods, or 3 ms. Choose $t_{high} = 2.2$ ms; then 2.2 ms $= 1.1R_A \times 0.1 \times 10^{-6}$ F; $R_A = 20$ kΩ.

13-6.4 Missing Pulse Detector

Transistor Q is added to the 555 one-shot in Fig. 13-14(a) to make a missing pulse detector. When E_i is at ground potential (0 V), the emitter diode of transistor Q clamps capacitor voltage V_C to a few tenths of a volt above ground. The 555 is forced into its idle state with a high output voltage V_o at pin 3. When

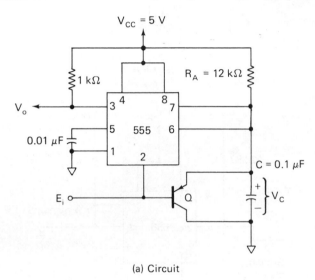

(a) Circuit

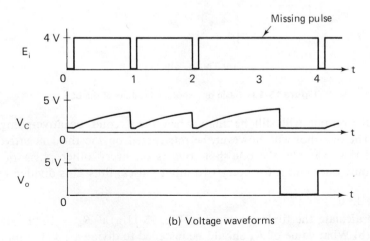

(b) Voltage waveforms

Figure 13-14 Missing pulse detector.

E_i goes high, the transistor cuts off and capacitor C begins to charge. This action is shown by the waveshapes in Fig. 13-14(b). If E_i again goes low before the 555 completes its timing cycle, the voltage across C is reset to about 0 V. If, however, E_i does *not* go low before the 555 completes its timing cycle, the 555 enters its normal state and output V_o goes low. This is exactly what happens if the $R_A C$ timing interval is slightly longer than the period of E_i and E_i suddenly misses a pulse. This type of circuit can detect a missing heartbeat. If E_i pulses are generated from a rotating wheel, this circuit tells when the wheel speed drops below a predetermined value. Thus the missing pulse detector circuit also performs speed control and measurement.

13-7 INTRODUCTION TO COUNTER TIMERS

When a timer circuit is connected as an oscillator and is used to drive a counter, the resultant circuit is *a counter timer*. Typically, the counter has many separate output terminals. One output terminal gives one pulse for each period T of the oscillator. A second output terminal gives one output pulse for every two periods ($2T$) of the oscillator. A third output terminal gives one output pulse for every four oscillator periods ($4T$), and so on, depending on the design of the counter. Thus each output terminal is rated in terms of the basic oscillator period T.

Some counters are designed so that their outputs can be connected together. The resultant output pulse is the *sum* of the individual output pulses. For example, if the first, second, and third output terminals are wired together, the result is one output pulse for every $1T + 2T + 4T = 7T$ oscillator periods. A counter with this capability is said to be *programmable*, because the user can program the counter to give one output pulse for any combination of timer outputs. One such programmable timer/counter is Exar's XR 2240. This integrated-circuit device is representative of the timer/counter family and some of its features will be studied next.

13-8 THE XR 2240 PROGRAMMABLE TIMER/COUNTER

13-8.1 Circuit Description

As shown in Fig. 13-15 and Appendix 5, the XR 2240 consists of one modified 555 timer, one 8-bit binary counter, and a control circuit. They are all contained in a single 16-pin dual-in-line package.

A positive-going pulse applied to *trigger* input 11 starts the 555 time base oscillator.

A positive-going pulse on *reset* pin 10 stops the 555 time base oscillator. The threshold voltage for both trigger and reset terminals is about $+1.4$ V.

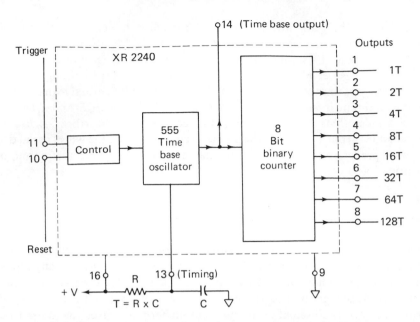

Figure 13-15   Block diagram of the XR 2240 programmable timer/counter.

The time base period T for one cycle of the 555 oscillator is set by an external RC network connected to the *timing* pin 13. T is calculated from

$$T = RC \tag{13-10}$$

where R is in ohms, C in farads, and T in seconds. R can range from 1 kΩ to 10 MΩ and C from 0.005 μF to 1000 μF. Thus, the period of the 555 can range from microseconds to hours.

Output of the 555 time base oscillator is available for measurement at pin 14 and also drives the 8-bit binary counter. Operation of the counter is discussed in Section 13-8.2.

13-8.2   Counter Operation

A simplified schematic of the 8-bit binary counter is shown in Fig. 13-16. Output of the 555 time base oscillator is shown as a switch. One side of the switch is connected to ground while the other side is wired to a 20-kΩ pull-up resistor. A regulated plus voltage is available at pin 15. Each negative-going edge from the 555 steps the 8-bit counter up by one count.

Normally, the 2240 is in its *reset* condition. That is, all 8 output pins (pins 1 to 8) act like open circuits, as shown by the output switch models in Fig. 13-16. Pull-up resistors (10 kΩ) should be installed, as shown, to those terminals that are going to be used. Outputs 1 and 4 will then be high in the reset condition.

When the 2240 is triggered (pulse applied to pin 11), all output switches of the counter are closed by the control circuit and outputs 1 to 8 go low. Thus,

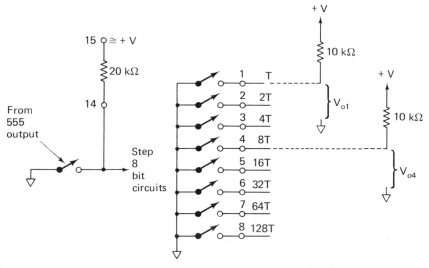

(a) Simplified outputs of the 2240

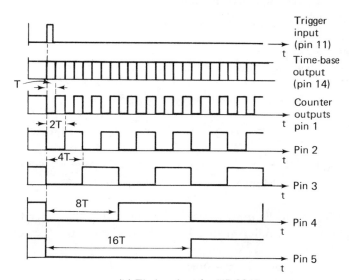

(b) Timing chart for XR 2240 outputs

Figure 13-16   Counter operation.

the counter begins its count with all outputs essentially grounded. At the end of every time base period, the 555 steps the counter once. The counter's *T* switch on terminal 1 opens after the first time base period (output 1 goes high) and closes after the second time base period. This counting action of the timer is shown in Fig. 13-16(b).

Table 13-1   Output terminal time chart

Terminal number	Time output stays low after trigger pulse
1	T
2	$2T$
3	$4T$
4	$8T$
5	$16T$
6	$32T$
7	$64T$
8	$128T$

Output pin 2 is labeled $2T$ in Fig. 13-16(a). It is seen from Fig. 13-16(b) that the output on pin 2 has stayed low for two time base periods ($2T$). Thus, the second output stays low for twice the time interval of the first output. This conclusion may be generalized to all outputs of the binary counter; that is, each output stays low for twice the time interval of the preceding output. Time intervals for pins 1 to 5 are shown in Fig. 13-16(b) and are given for all outputs in Table 13-1.

Example 13-7

After triggering, how long will the following output terminals stay low? (a) Pin 3; (b) pin 4; (c) pin 7; (d) pin 8. $R = 100$ kΩ and $C = 0.01$ μF.
Solution. By Eq. (13-10), the time base period is

$$T = (100 \times 10^3)(0.01 \times 10^{-6}) = 1 \text{ ms}$$

From Table 13-1, (a) $t_{\text{low}} = 4(1 \text{ ms}) = 4$ ms; (b) $t_{\text{low}} = 8(1 \text{ ms}) = 8$ms; (c) $t_{\text{low}} = 64(1 \text{ ms}) = 64$ ms; (d) $t_{\text{low}} = 128(1 \text{ ms}) = 128$ ms.

The conclusion to be drawn from Example 13-7 is that after triggering, there are eight pulses of different time intervals available from the counter timer.

13-8.3 Programming the Outputs

The output circuits are designed to be used either individually or wired together, which is called *wire-or*. The term wire-or means two or more output terminals can be jumpered together with a common wire (output bus) to a single pull-up resistor, as shown in Fig. 13-17(a). The resultant timing cycle for V_o is found by redrawing the individual timing of pins 4 and 5 in Fig. 13-17(b). Here we see that as long as either pin 4 *or* pin 5 is low, V_o will be low. Only when both outputs go high (output switches open) will the output go high. Thus

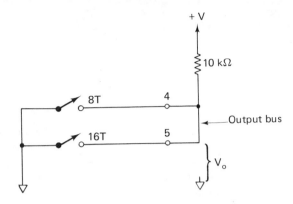

(a) Pins 4 and 5 are wired together to program 24T

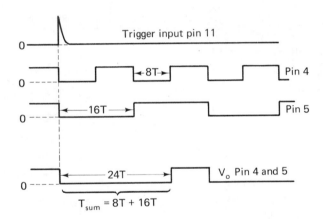

(b) Common bus V_o stays low as long as either pin 4 or pin 5 stays low

Figure 13-17 Programming the outputs.

the timing cycle for the output bus is found simply by calculating the sum, T_{sum}, of the individual outputs.

Example 13-8

Calculate the timing cycle for (a) Fig. 13-17(a); (b) a circuit where pins 3, 6, and 7 are jumpered to a common bus. Let $T = 1$ s.

Solution. (a) $T_{sum} = 8T + 16T = 24T = 24 \times 1$ s $= 24$ s;   (b) $T_{sum} = 4T + 32T + 64T = 100T = 100 \times 1$ s $= 100$ s.

By using switches instead of jumper wires, T_{sum} can be easily changed or *programmed* for any desired timing cycle from T to $255T$.

13-9 TIMER/COUNTER APPLICATIONS

13-9.1 Timing Applications

The 2240 is wired for monostable operation in the programmable timer application of Fig. 13-18. When the trigger input goes high, the output bus goes low for a timing cycle period equal to T_{sum} (see Section 13-8.3). At the end of the timing cycle, the output bus goes high. The connection from output bus via a 51-kΩ resistor to reset pin 10 forces the timer to reset itself when the output bus goes high. Thus after each trigger pulse, the 2240 generates a timing interval selected by the program switches.

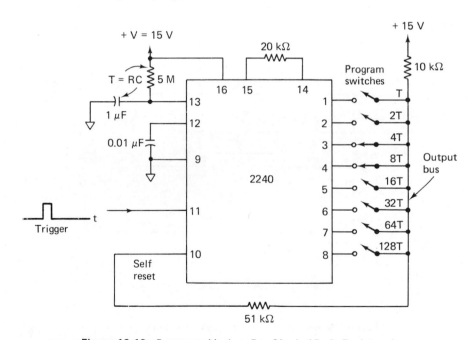

Figure 13-18 Programmable timer 5 to 21 min 15 s in 5-s intervals.

Example 13-9

In Fig. 13-18, $C = 1.0$ μF and $R = 5$ MΩ to establish a time base period given by Eq. (13-10) to be 5 s. What is (a) the timing cycle for switch positions shown in Fig. 13-18; (b) the minimum programmable timing cycle; (c) the maximum programmable timing cycle?

Solution. (a) $T_{sum} = 4T + 8T = 12T = 12 \times 5$ s $= 60$ s $= 1$ min; (b) minimum timing cycle is $1T = 5$ s; (c) with all program switches closed,

$$T_{\text{sum}} = T + 2T + 4T + 8T + 16T + 32T + 64T + 128T = 225T$$

$$225T = 225 \times 5\,\text{s} = 1275\,\text{s} = 21\,\text{min}\ 15\,\text{s}$$

13-9.2   Free-Running Oscillator, Synchronized Outputs

The 2240 operates as a free-running oscillator in the circuit of Fig. 13-19. The reset terminal is grounded so that the 2240 will stay in its timing cycle once it is started. When power is applied, R_R and C_R couple a positive-going pulse into trigger input 11 to start the internal time base oscillator running.

Each output is wired through an external control switch to an individual pull-up resistor. A square-wave output voltage is available at each counter output. Their frequencies have a binary relationship. That is, the frequency available at each pin is one-half the frequency present at the preceding pin. The waveshapes are identical to those in Fig. 13-16(b). Observe that the *period* of the f_1 frequency at pin 1 is twice the time base period rating T or $2(T)$. Thus $f_1 = 1/2T$. At pin 4, the period is $2(8T)$ and $f_4 = 1/16T$.

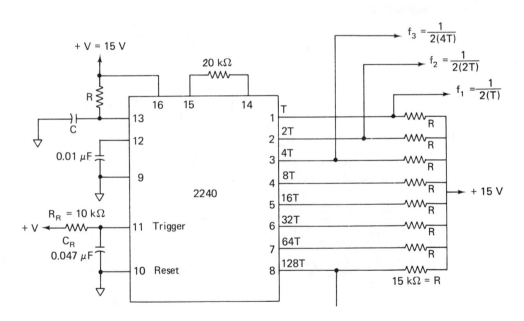

Figure 13-19   Free-running oscillator with synchronized outputs.

Example 13-10

In Fig. 13-19, $T = 2.5$ ms; what frequencies are present at (a) output 1; (b) output 2; (c) output 3; (d) output 4?
Solution. Tabulating calculations, we obtain:

Pin number	Time base rating	Period	Frequency (Hz)
1	T	$2T = 5$ ms	200
2	$2T$	$4T = 10$ ms	100
3	$4T$	$8T = 20$ ms	50
4	$8T$	$16T = 40$ ms	25

The connections to pins 10 and 11 may be removed to allow the oscillator to be started with a positive-going trigger pulse at pin 11. To stop oscillation, apply a positive-going pulse to reset pin 10.

13-9.3   Binary Pattern Signal Generator

Pulse patterns similar to those shown in Fig. 13-20 are generated by a modified version of Fig. 13-19. The modification requires the eight output resistors to be replaced by program switches and a single 10-kΩ resistor similar to that shown in Fig. 13-18. Also eliminate the 51-kΩ resistor between the output bus and the self-reset terminal.

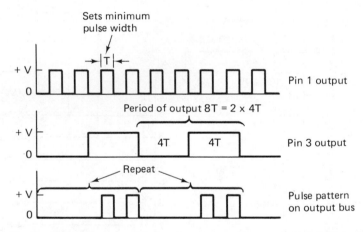

Figure 13-20   Binary pattern signal gererator with outputs T and $4T$ connected to output bus.

The output is a train of pulses (as shown in Appendix 5, p. 000) that depends on which program switches are closed. The period of the pulse pattern is set by the highest program switch that is closed, and the pulse width is set by the lowest program switch that is closed. For example, if the $4T$ (pin 3) and $1T$ (pin 1) switches are closed, the pulse pattern is repeated every $2 \times 4T = 8T$ seconds (see Fig. 13-20). The minimum pulse width is $1T$. To determine the actual pulse pattern, refer to the timing chart in Fig. 13-16 (b). If switches $1T$ and $4T$ are closed, there is an output pulse only when there are high output pulses from each line. The repeating pulse patterns are shown in Fig. 13-20.

13-9.4 Frequency Synthesizer

The output bus in Fig. 13-21(a) is capable of generating any one of 255 related frequencies. Each frequency is selected by closing the desired program switches to program a particular frequency at output V_o.

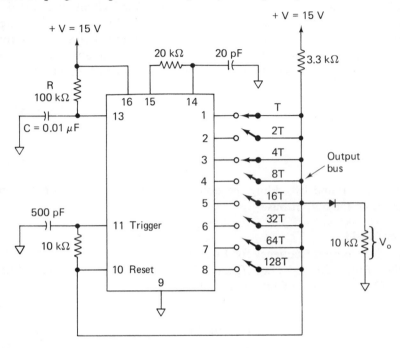

(a) Frequency synthesizer connections

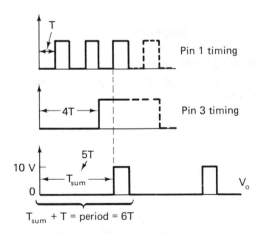

(b) Output voltage for program switches 1 and 4 closed

Figure 13-21   Frequency synthesizer, $T = 1$ ms, $f = 166$ Hz.

To understand circuit operation, assume the output bus goes high. This will drive reset pin 10 high and couple a positive-going pulse into trigger pin 11. The reset terminal going positive resets the 2240 (all outputs low). The positive pulse on pin 10 retriggers the 2240 time base oscillator, to begin generation of a time period that depends on which program switches are closed. For example, assume that switches T and $4T$ are closed in Fig. 13-21(a). The timing for these switches is shown in Fig. 13-21(b). The output bus stays low for $4T$ from pin 3 plus $1T$ from pin 1 before going high (to initiate a reset-retrigger sequence noted above). The time period and frequency of the output signal V_o is thus expressed by

$$\text{period} = (T_{\text{sum}} + T) \qquad (13\text{-}11a)$$

and

$$f = \frac{1}{\text{period}} \qquad (13\text{-}11b)$$

where T_{sum} is found by adding the time base rating for each output terminal connected to the output bus.

Example 13-11

Find the output frequency for Fig. 13-21(a).
Solution. From Eq. (13-11a), $T_{\text{sum}} = 1T + 4T = 5T$, and period $= (T_{\text{sum}} + T) = 6T = 6 \times 1$ ms $= 6$ ms. By Eq. (13-11b),

$$f = \frac{1}{6 \times 10^{-3}\text{ s}} \doteq 166 \text{ Hz}$$

Other representative applications of the XR 2240 are shown in Appendix 5.

PROBLEMS

13-1. What are the operating modes of the 555 timer?

13-2. In Fig. 13-6(a), $R_A = R_B = 10$ kΩ, $C = 0.1$ μF. Find (a) t_{high}; (b) t_{low}; (c) frequency of oscillation.

13-3. Using the graph of Fig. 13-7, estimate the free-running frequency of oscillation f if $(R_A + 2R_B) = 1$ MΩ and $C = 0.02$ μF.

13-4. What is the duty cycle in Problem 13-2?

13-5. In Example 13-1, R_A and R_B are increased by a factor of 10 to 68 kΩ and 33 kΩ. Find the new frequency of oscillation.

13-6. In Fig. 13-8, R_A and R_B are each reduced to 5 kΩ. What is the effect on (a) the duty cycle; (b) the period T of the output?

13-7. In Fig. 13-9, at what value should the 10-kΩ resistor be set for a 2-kHz output from the B 555?

13-8. What is the duty cycle for Fig. 13-10 if the 10-kΩ potentiometer is set in the middle (5 kΩ on either side of the wiper)?

13-9. In Fig. 13-9(a), $R_A = 100$ kΩ and $C = 0.1$ μF. Find t_{high}.

13-10. R_A is changed to 20 kΩ in Example 13-5. Find t_{high}.

13-11. In Example 13-6(b), what value of R_A is required to divide a 1-kHz signal by 2?

13-12. Refer to Example 13-7, how long will the following output terminals stay low? (a) Pin 1; (b) pin 2; (c) pin 5; (d) pin 6.

13-13. In Fig. 13-17(a), T is set for 1 ms and pins 2, 4, 6, and 8 are connected to the output bus. Find the timing interval.

13-14. In Problem 13-13, the odd-numbered pins, 1, 3, 5, and 7, are connected to the output bus. Find the timing interval.

13-15. In Example 13-9, C is changed to 0.1 μF and R to 500 kΩ. Find (a) the time base period; (b) the timing cycle for switch positions shown in Fig. 13-18; (c) the maximum timing cycle.

13-16. In Example 13-10, what frequencies are present at pins (a) 5; (b) 6; (c) 7; (d) 8?

13-17. In Fig. 13-21, only switches to pins 1, 2, 3, and 4 are closed. Find the output frequency.

power supplies
and power amplifiers

14

14-0   INTRODUCTION

Most electronic devices require dc voltages to operate. Batteries are useful in low-power or portable devices, but operating time is limited unless the batteries are recharged or replaced. The most readily available source of power is the 60-Hz 110-V ac wall outlet. The circuit that converts this ac voltage to a dc voltage is called a *dc power supply*.

The most economical dc power supply is some type of rectifier circuit. Unfortunately, some ac ripple voltage rides on the dc voltage, so the rectifier circuit does not deliver pure dc. An equally undesirable characteristic is a reduction in dc voltage as more load current is drawn from the supply. Since dc voltage is *not* regulated (that is, constant with changing load current), this type of power supply is classified as *unregulated*. Unregulated power supplies are introduced in Sections 14-1 and 14-2. It is necessary to know their limitations before such limitations can be minimized or overcome by adding regulation.

A circuit that essentially eliminates variation of power supply voltage with changing load currents is called a *voltage regulator*. Normally, all the ac ripple voltage is also eliminated by the regulator circuit. When a voltage regulator is connected to a rectifier circuit, the result is a *regulated power supply*.

Voltage regulators with outstanding performance can be made quickly and easily by using the op amp. An even greater benefit provided by the op amp is the ease of adding options such as short-circuit protection of the regulated power supply. With a few minor changes and at a very nominal cost, the regulator can be converted to a power amplifier that rivals high fidelity amplifiers in performance.

The growing use of op amps in making voltage regulators, together with their associated zener references (Sections 14-5 and 14-6), boost transistor (Section 14-7), and current limiting (Section 14-8), led to the development of *integrated voltage circuit regulators.* Manufacturers have crammed all of these features into a *single* chip, plus additional features such as thermal protection against excessive heat and safe area protection against current, voltage, or power overloads.

There is a bewildering array of modern IC regulators. You can buy fixed positive voltage regulators of 5-, 6-, 8-, 12-, 15-, 18-, and 24-V output. You can also buy fixed negative voltage regulators to output -5, -6, -8, -12, -15, -18, and -24 V.

The tolerance on the fixed regulators is usually $\pm 5\%$. Suppose that you need precisely 5.00 V or a value that is not available as a fixed regulator; or that you want a regulator that you can adjust to 5 V for testing TTL logic and then adjust to 12 V to test a car radio or charge its battery. The ingenious IC manufacturers have created adjustable positive voltage regulators and also adjustable negative voltage regulators. In addition they make fixed or adjustable dual-polarity tracking regulators.

You can select current output capabilities of 0.1 to 5.0 A. The current output can be increased by the addition of boost transistors. We have selected three IC regulators that are representative. An adjustable positive regulator and an adjustable negative regulator are presented in Section 14-9. A dual-polarity tracking regulator is presented in Section 14-10. We begin with the circuitry that feeds these regulators—the unregulated power supply.

14-1 INTRODUCTION TO THE UNREGULATED POWER SUPPLY

14-1.1 Power Transformer

A transformer is required for reducing the 110-V ac wall current to the lower ac value required by transistor, ICs, and other electronic devices. Transformer voltages are given in terms of rms values. In Fig. 14-1, the transformer is rated as 110 V to 24 V center tap. With 110 V rms connected to the primary, 24 V rms is developed between secondary terminals 1 and 2. A third lead, brought out from the center of the secondary, is called a *center tap*, CT. Between terminals CT and 1 or CT and 2, the rms voltage is 12 V.

An oscilloscope would give the sinusoidal voltages shown in Fig. 14-1. The maximum instantaneous voltage E_m is related to the rms value E_{rms} by

$$E_m = 1.4(E_{rms}) \tag{14-1}$$

In Fig. 14-1(a), voltage polarities are shown for the positive half-cycle; those for the negative half-cycle are shown in Fig. 14-1(b).

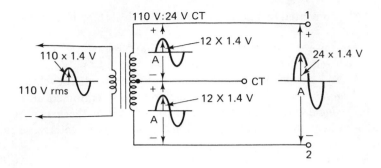

(a) Peak voltages for the positive half-cycle

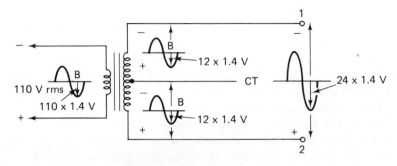

(b) Peak voltages for the negative half-cycle

Figure 14-1   Power transformer.

Example 14-1

Find E_m in Fig. 14-1 between terminals 1 and 2.
Solution. By Eq. (14-1), $E_m = 1.4(24 \text{ V}) = 34 \text{ V}$.

14-1.2   Rectifier Diodes

In Fig. 14-2(a), four diodes are arranged in a diamond configuration called a *full-wave bridge rectifier*. They are connected to terminals 1 and 2 in the transformer of Fig. 14-1. When terminal 1 is positive with respect to terminal 2, diodes D_1 and D_2 conduct. When terminal 2 is positive with respect to terminal 1, diodes D_3 and D_4 conduct. The result is a pulsating dc voltage between the output terminals.

14-1.3   Filter Capacitor

The pulsating dc voltage in Fig. 14-2(a) is not pure dc, so a filter capacitor is placed across the dc output terminals, as shown in Fig. 14-2(b). This capacitor smooths out the pulsations and gives an almost pure dc output voltage, V_L. V_L

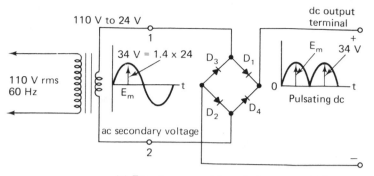

(a) Transformer and four diodes reduce 162 V peak ac
from wall outlet to 34 V peak pulsating dc.

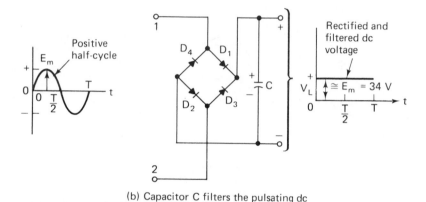

(b) Capacitor C filters the pulsating dc
in (a) to give a dc load voltage

Figure 14-2 Transformer plus rectifier diodes plus filter capacitor equals
unregulated power supply.

is the unregulated voltage that supplies power to the load. The filter capacitor
is typically a large electrolytic capacitor, 500 μF or more.

14-1.4   The Load

In Fig. 14-2(b), nothing other than the filter capacitor is connected across
the dc output terminals. The unregulated power supply is said to have no load.
This means that the *no-load current*, or 0 load current, I_L, is drawn from the
output terminals. Usually, the maximum expected load current, or full-load
current, to be furnished by the supply is known. The load is modeled by resistor
R_L as shown in Fig. 14-3(a). As stated in Section 14-0, the load voltage changes
as the load current changes in an unregulated power supply. The manner in
which this occurs is examined next.

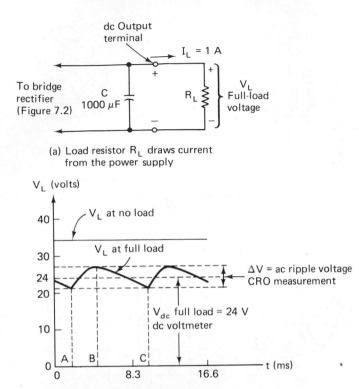

(a) Load resistor R_L draws current
     from the power supply

(b) Load voltage changes from 34 V at no
     load to 24 V plus ripple at full load

Figure 14-3   Variation of dc load voltage and ac ripple voltage from no-load
current to full-load current.

14-2   PREDICTING DC VOLTAGE REGULATION
AND AC RIPPLE VOLTAGE

14-2.1   Load Voltage Variations with Load Current

A dc voltmeter connected across the output terminals in Fig. 14-2(b)
measures the no-load voltage, or

$$V_{\text{dc no load}} = E_m$$

From Example 14-1, $V_{\text{dc no load}}$ is 34 V. An oscilloscope would also show the
same value with no ac ripple voltage, as in Fig. 14-3(b). Now suppose that a
load R_L was connected to draw a full-load dc current of $I_L = 1$ A, as in Fig.
14-3(a). An oscilloscope now shows that the load voltage V_L has a lower *average*,
or dc value V_{dc}. Moreover, the load voltage has an ac ripple component, ΔV_o,
superimposed on the dc value. The average value measured by a dc voltmeter
is 24 V. The peak-to-peak ripple voltage is $\Delta V_o = 5$ V.

There are two conclusions to be drawn from Fig. 14-3(b). First, the dc load voltage goes down as dc load current goes up; how much the load voltage drops can be estimated by a simple technique explained in Section 14-2.2. Second, the ac ripple voltage increases from 0 V at no-load current to a large value at full-load current. As a matter of fact, the ac ripple voltage increases directly with an increase in load current. The amount of ripple voltage can also be estimated, by a technique explained in Section 14-2.3.

14-2.2 DC Voltage Regulation Curve

In the unregulated power supply circuit Fig. 14-4(a), the load R_L is varied so that we can record corresponding values of dc load current and dc load voltage. The dc meters respond only to the average (dc) load current or voltage. If corresponding values of current and voltage are plotted, the result is the *dc voltage regulation curve* of Fig. 14-4(b). For example, point 0 represents the no-load condition, $I_L = 0$ and $V_{dc\ no\ load} = E_m = 34$ V. Point A represents the full-load condition, $I_L = 1$ A and $V_{dc\ full\ load} = 24$ V.

There is a general procedure to *estimate* the value of dc load voltage for any load current. These values depend primarily on the transformer rating, provided that the filter capacitor C is greater than 200 μF. Power supply transformers are rated by rms voltage ratios and maximum secondary load current I_T. For example, the inexpensive transformer of Fig. 14-4(a) is rated as 110 V: 24 V at 1 A. Find the ratio of your load current I_L to I_T and locate this point on the horizontal axis of Fig. 14-5. Proceed vertically to the estimating curve and then horizontally to read V_{dc} on the vertical axis as a fraction of E_m. The procedure is illustrated by an example.

Example 14-2

In Fig. 14-4, $E_m = 24$ V $\times$ 1.4 $= 34$ V. $I_T = 1$ A. Find the dc load voltage V_{dc} at (a) $I_L = 0.5$ A; (b) $I_L = 1.0$ A.
Solution. (a) $I_L/I_T = 0.5$ A$/1.0$ A $= 0.5$ Locate point M in Fig. 14-5 and read

$$V_{dc} = 0.85E_m = 0.85 \times 34 = 29\ V$$

(b) $I_L/I_T = 1.0$ A$/1.0$ A $= 1.0$. Locate the maximum load point where $V_{dc} = 0.7E_m = 0.7 \times 34 = 24$ V.

If a transformer were rated for 110 V: 24 V and $I_T = 2$ A, V_{dc} would be approximately 31 V and 29 V for parts (a) and (b) of Example 14-2.

14-2.3 Estimating and Reducing AC Ripple Voltage

It was concluded in Section 14-2.1 and Fig. 14-3(b) that ac ripple increases as dc load current increases. The worst case of ac ripple occurs at maximum load current. The peak-to-peak ac ripple voltage ΔV can be estimated from Fig. 14-6 as shown by the following example.

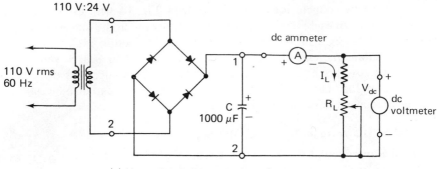

(a) Unregulated power supply performance
measured with dc ammeter and voltmeter

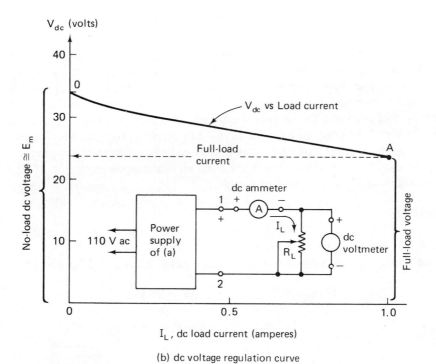

(b) dc voltage regulation curve

Figure 14-4   Dc load voltage varies with load current in (a) as shown by
the voltage regulation curve in (b).

Example 14-3

What is ΔV for the power supply of Fig. 14-4 if the load current is 1 A and
(a) $C = 1000 \ \mu F$; (b) $C = 2000 \ \mu F$; (c) $C = 5000 \ \mu F$?
Solution. In Fig. 14-6, enter the horizontal axis at $I_L = 1.0$ A. Proceed vertically
to intersections A, B, and C with curves respectively labeled $C = 1000 \ \mu F$,

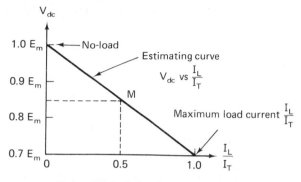

Figure 14-5   Estimating dc load voltage.

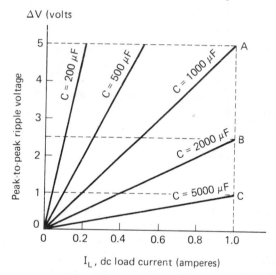

Figure 14-6   Estimating peak-to-peak ac ripple voltage ΔV.

$C = 2000\ \mu$F, and $C = 5000\ \mu$F. Read ΔV from the vertical axis as shown to get (a) $\Delta V = 5$ V; (b) $\Delta V = 2.5$ V; (c) $\Delta V = 1$ V.

Conclusion

From Fig. 14-6, ac ripple is shown to be reduced by increasing the value of filter capacitor C. Doubling C halves the ripple. If an oscilloscope is not available, an ac voltmeter can be connected across the load R_L to measure the approximate rms ripple voltage V_r. V_r is related to ΔV by

$$V_r \cong \frac{\Delta V}{3} \qquad (14\text{-}2)$$

Example 14-4

Find the ac voltmeter readings for the peak-to-peak ripple voltages in Example 14-3.

Solution. (a) $V_r = 5\,\text{V}/3 = 1.7\,\text{V}$;   (b) $V_r = 2.5\,\text{V}/3 = 0.8\,\text{V}$;   (c) $V_r = 1\,\text{V}/3 = 0.3\,\text{V}$.

14-2.4   Minimum Instantaneous Load Voltage

Refer to Fig. 14-3(b) and note that ΔV is centered on the dc load voltage V_{dc}. The minimum *instantaneous* value of load voltage V_L will be

$$\text{minimum } V_L = V_{\text{dc}} - \frac{\Delta V}{2} \tag{14-3}$$

Since V_{dc} goes down and ΔV goes up with increasing load current, the lowest minimum value of V_L occurs at full load current. This value of V_L places a limit on any voltage regulator that will be attached to the unregulated power supply. But before we proceed to the voltage regulator, another type of unregulated power supply will be studied so that it can be used with the power amplifier.

14-3   BIPOLAR AND TWO-VALUE UNREGULATED POWER SUPPLIES

14-3.1   Bipolar or Positive and Negative Power Supplies

Many electronic devices need both positive $(+)$ and negative $(-)$ voltages. These voltages are measured with respect to a third common (or grounded) terminal. To obtain a positive and negative voltage, either two secondary transformer windings or one center-tapped secondary winding is needed.

A transformer rated at 110 V: 24 V CT is shown in Fig. 14-7. Diodes D_1 and D_2 make terminal 1 positive with respect to center tap CT. Diodes D_3 and D_4 make terminal 2 negative with respect to the center tap. From Eq. 14-1 and Section 14-2.1, both no-load dc voltages are $1.4 \times 12\,\text{V rms} = 16\,\text{V}$. Capacitors $C+$ and $C-$, respectively, filter the positive and negative supply

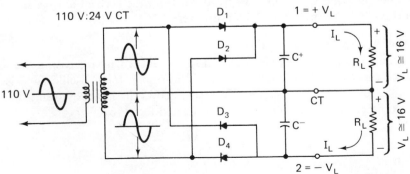

Figure 14-7   Bipolar power supply.

voltages. As outlined in Figs 14-5 and 14-6, the ac ripple voltage and dc voltage regulation may be predicted for both load voltages.

14-3.2 Two-Value Power Supplies

If the center tap of the power supply of Fig. 14-7 is grounded, we have a *bipolar* power supply. It is shown schematically in Fig. 14-8(a). If terminal 2 is grounded as in Fig. 14-8(b), we have a two-value positive supply. Finally, by grounding terminal 1 in Fig. 14-8(c), we get a two-value negative power supply. This indicates the versatility of the center-tapped transformer. The bipolar power supply is used with an audio amplifier in Section 14-11.

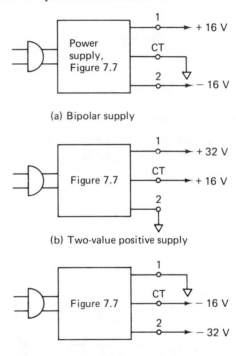

(a) Bipolar supply

(b) Two-value positive supply

(c) Two-value negative supply

Figure 14-8 Bipolar and two-value power supplies.

14-4 NEED FOR VOLTAGE REGULATION

Previous sections have shown that the unregulated power supply has two undesirable characteristics: the dc voltage decreases and the ac ripple voltage increases as load current increases. Both disadvantages can be minimized by adding a voltage-regulator section to the unregulated supply as in Fig. 14-9. The resulting power supply is classified as a *voltage-regulated supply*. Before turning to the op amp for an almost ideal voltage regulator, we must first learn something about an element that gives a stable reference voltage, the zener

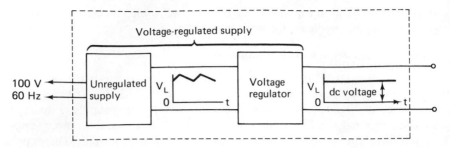

Figure 14-9 Unregulated supply plus a voltage regulator gives a voltage-regulated power supply.

diode. The zener diode and its applications as a basic voltage regulator are studied in Section 14-5.

14-5 ZENER DIODE REGULATOR

14-5.1 Characteristics of the Zener Diode

The symbol for a zener diode is shown in Fig. 14-10. When forward-biased (with the + terminal of E_i connected to the arrowhead), the zener behaves as any silicon diode does. Its terminal voltage stays at about 0.6 to 0.7 V no matter what current it conducts. However, let a reverse-biased voltage (with the −

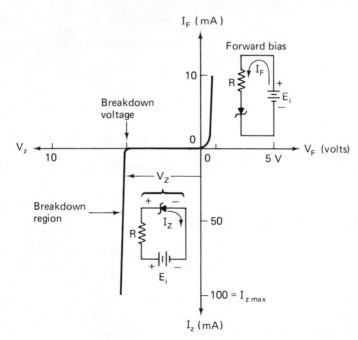

Figure 14-10 Zener diode characteristics.

terminal of E_i connected to the arrowhead) be connected across the zener. As the reverse-biased voltage is increased, the zener voltage will increase until a zener *breakdown voltage* V_z is reached. For all values of E_i greater than V_z the zener's terminal voltage stays constant no matter what current flows through the zener. As shown in Fig. 14-10, this characteristic of the zener is called the *breakdown region*. Zener diodes are designed to operate in the breakdown region. They are available with zener voltage ratings from a few volts to a few hundred volts. In Fig. 14-10, $V_z = 5$ V.

14-5.2 Zener Voltage Regulator

If a load is connected across a zener as in Fig. 14-11(a), the load voltage equals the zener voltage. As the load current varies in Fig. 14-11(b), the load voltage stays at V_z. What happens is that the zener absorbs all current not drawn by the load. We shall not go into the analysis and design of zener regulators. Our objective is to reach the following conclusions.

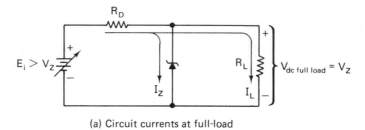

(a) Circuit currents at full-load

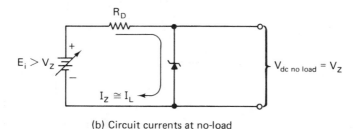

(b) Circuit currents at no-load

Figure 14-11 Zener voltage regulator.

1. Once the zener is reverse-biased, its terminal voltage stays constant at V_z. V_z is set by the zener one buys.
2. The zener needs to be reverse-biased into breakdown to set up our reference voltage. To ensure breakdown, the zener must conduct a few mA, I_{zon}, typically 5 mA. This current is set by resistor R_D, V_z, and the input voltage E_i according to

$$I_{zon} = \frac{E_i - V_z}{R_D} \cong 5 \text{ mA} \qquad (14\text{-}4)$$

14-6 BASIC OP AMP VOLTAGE REGULATOR

14-6.1 Op Amp Regulator

Close examination of the op amp regulator in Fig. 14-12(a) shows that it is the noninverting amplifier of Section 3-6. The load voltage V_o is set by the battery reference voltage V_{ref} and feedback resistors R_f and R_i according to

$$V_o = \frac{R_f + R_i}{R_i} V_{ref} \qquad (14\text{-}5)$$

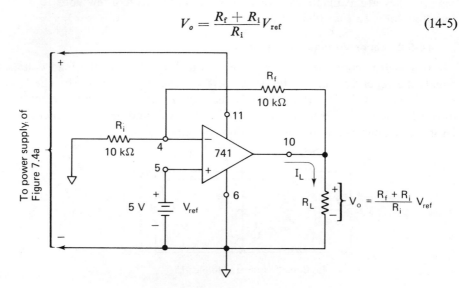

(a) Basic op amp regulator

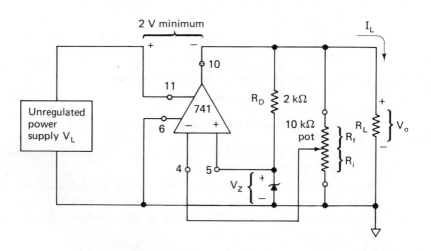

(b) Practical op amp voltage-regulated power supply

Figure 14-12 Basic op amp regulators.

As load current I_L varies, V_o is held constant according to Eq. (14-5). The ripple voltage and changing dc voltage of the unregulated supply are absorbed by the op amp. Voltage between pins 11 and 6 should not exceed 36 V. Output pin 10 and input pin 5 must always be at least 2 V below pin 11; otherwise, the op amp will go into negative or positive saturation, respectively.

Either resistor R_i or R_f or the battey V_{ref} may be located away from the op amp and varied to change the dc output voltage V_o. Such an arrangement is called *remote programming*. The op amp circuit now regulates the load voltage to be constant and independent of the load current.

14-6.2   Op Amp Regulator with Zener Reference

In Fig. 14-12(b), the battery is replaced by a zener diode and bias resistor R_D. R_D is determined by rearranging Eq. (14-4). Thus the voltage at pin 10 must always be a few volts more positive than pin 5 to keep the zener diode operating in its breakdown region.

Pin 11 must always be about 2 V more positive than pin 10 to avoid op amp saturation. The 2-V difference between minimum unregulated supply voltage V_L and regulated supply voltage V_o is so important that it is specified for all IC regulated power supplies as *minimum input-output differential voltage*. These points are illustrated by the following design procedure.

To design Fig. 14-12(b) as a 15 V voltage regulator:

1. Select a zener about $\frac{1}{3}$ to $\frac{1}{2}$ of V_o or $V_z = 5$ V.
2. Calculate R_D from Eq. (14-4):

$$R_D = \frac{15\text{ V} - 5\text{ V}}{5\text{ mA}} 2\text{ k}\Omega$$

3. Pick $R_f + R_i$ to draw about 1 mA from the regulated output $V_o = 15$ V or $R_f + R_i = 15$ V/1 mA $\cong$ 15 kΩ; select a 10-kΩ potentiometer.
4. From Eq. 14-5, solve for R_i:

$$R_i = (R_f + R_i)\frac{V_{ref}}{V_o} = 10\text{ k}\Omega\left(\frac{5\text{ V}}{15\text{ V}}\right) = 3.33\text{ k}\Omega$$

Set the pot for $R_i = 3.33$ kΩ or install fixed resistors of $R_i = 3.33$ kΩ and $R_f = 10$ k$\Omega - 3.33$ k$\Omega = 6.67$ kΩ.

5. The minimum unregulated supply voltage (voltage at pin 11) must be $V_o + 2$ V = 17 V.

14-7   CURRENT-BOOSTING THE OP AMP REGULATOR

14-7.1   Pass Transistor

The op amp regulator of Fig. 14-12 can only furnish load currents of up to about 5 mA. By adding a pass transistor as in Fig. 14-13(a), the maximum load current capability of the voltage regulator is extended by a factor of

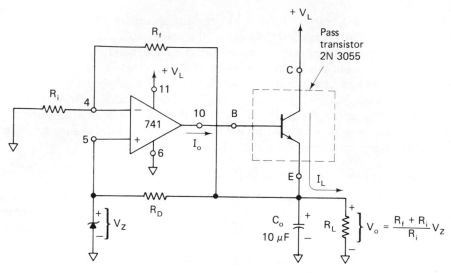

(a) Voltage regulator with pass transistor

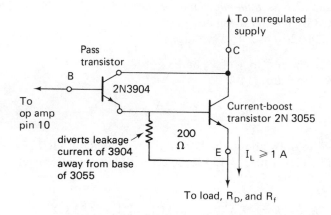

(b) Adding a current-boost transistor for large load currents

Figure 14-13 Increasing the load-current capability of a basic op amp regulator.

approximately 100 to 0.5 A. To see how this is accomplished, assume that V_o is set at 15 V according to Eq. (14-5) or the design procedure in Section 14-6.2. Then let R_L be 30 Ω so that the load current is $I_L = V_o/R_L = 15$ V/30 $\Omega = 0.5$ A. I_L is furnished by *emitter* terminal E of the pass transistor. If β (the current gain) of the transistor is 100, the transistor requires a base current drive of $I_L/\beta = 500$ mA/100 $= 5$ mA. The op amp can furnish 5 mA from its output terminal (pin 10) to the base terminal B.

Changes in unregulated supply voltage V_L due to either ac ripple or poor dc voltage regulation are absorbed by the collector of the pass transistor. The pass transistor should be heat-sinked to have a power rating P_D of

$$P_D = (V_L - V_o)I_{L \text{ full load}} \tag{14-6}$$

Example 14-5

If $V_L = 24$ V, $I_{L \text{ full load}} = 0.5$ A, and $V_0 = 15$ V in Fig. 14-13(a), find the power dissipation of the pass transistor.
Solution. From Eq. (14-6), $P_D = (24 - 15)$ V $\times$ 0.5 A = 4.5 W.

14-7.2 Current-Boost Transistor

The pass transistor of Fig. 14-13(a) can be replaced by two transistors for further increase in load current-capability. As shown in Fig. 14-13(b), the op amp can furnish up to 5 mA into the base of the pass transistor. The current gain of the pass transistor results in 100 $\times$ 5 mA or 0.5 A as the base current drive into the base of the current-boost transistor. The current gain of the current-boost transistor allows an emitter or load current that exceeds 1 A. In practice, the power dissipation limitation of the boost transistor limits load currents to a few amperes. For more current output, more current-boost transistors must be added in parallel with the original boost transistor. (See Driscoll and Coughlin, *Solid State Devices and Applications*, Prentice-Hall, Englewood Cliffs, N.J., 1975, p. 244.)

14-8 SHORT-CIRCUIT OR OVERLOAD PROTECTION

Short-circuit protection is easily incorporated into the voltage regulator by adding a current-limiting transistor Q_S and a current-sensing resistor R_S as shown in Fig. 4-14. The load current I_L flows through R_S. Usually, the regulator is designed to furnish up to a specified maximum load current, $I_{L \text{ max}}$. $I_{L \text{ max}}$ is large with respect to the currents through R_D and R_f.

Resistor R_S monitors the load current I_L to prevent the regulator from drawing too much load current. When I_L reaches $I_{L \text{ max}}$, the voltage drop across R_S rises to about 0.6 V. Transistor Q_S then turns on, and its collector conducts the op amp's output current away from the base of the 2N 3055 pass transistor. Once Q_S turns on, the emitter current of the pass transistor, and consequently the load current, is held constant at $I_{L \text{ max}}$. R_S sets the value of $I_{L \text{ max}}$ according to

$$I_{L \text{ max}} = \frac{0.6 \text{ V}}{R_S} \tag{14-7}$$

Example 14-6

Choose R_S for a maximum load current of 0.5 A for the regulator of Fig. 14-14.

Solution. Rearranging Eq. (14-7), we obtain

$$R_S = \frac{0.6\text{ V}}{I_{L\text{ max}}} = \frac{0.6\text{ V}}{0.5\text{ A}} = 1.2\ \Omega$$

The worst overload occurs when the output terminals are short-circuited. In this case, the pass transistor must be heat-sinked to be able to dissipate power equal to

$$P_D = V_L I_{L\text{ max}} \tag{14-8}$$

For example, if $V_L = 20$ V and $I_{L\text{ max}} = 0.5$ A, the transistor must dissipate 10 W.

The features of voltage programming, current boost, and overload protection are incorporated in *integrated circuit regulators*. IC voltage regulators are versatile, convenient to use, and relatively inexpensive. They are available in four classifications:

1. Positive voltage
2. Negative voltage

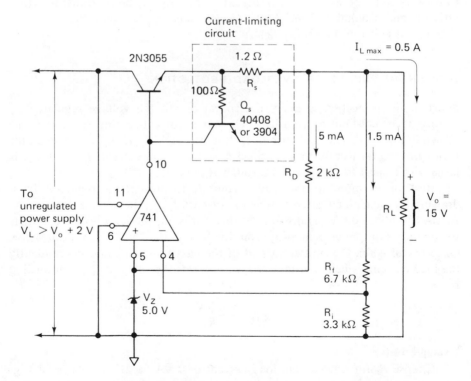

Figure 14-14 The current-limiting circuit holds I_L to a maximum of 0.5 A even if R_L is short-circuited.

3. Dual voltage

4. Special-purpose voltage regulators

Data sheets and application notes are available from manufacturers (at no cost) and show how to use the devices. Two IC regulators are selected from the vast number available to illustrate how to design and build your own regulator.

14-9 ADJUSTABLE THREE-TERMINAL POSITIVE-VOLTAGE REGULATOR—THE LM317HV

14-9.1 Adjusting the Regulated Output Voltage

The LM317HV adjustable, positive-voltage regulator has only three terminals, as shown in Fig. 14-15(a). Installation is simple, as shown in Fig. 14-15(b). The LM317 maintains exactly 1.2 V between its output and adjust terminals. A 240-Ω resistor, R_1, is connected between these terminals to conduct a current of 1.2 V/240 Ω = 5 mA. This 5 mA flows through R_2. If R_2 is adjustable, the voltage drop across it, V_{R2}, will equal $R_2 \times 5$ mA. Output voltage of the regulator is set by V_{R2} plus the 1.2-V drop across R_1. In general terms, V_o is given by

$$V_o = \frac{1.2\,\text{V}}{R_1}(R_1 + R_2) \tag{14-9a}$$

Normally, $R_1 = 240\ \Omega$. Thus any desired value of regulated output voltage is set by trimming R_2 to a value determined from

$$V_o = 1.2\,\text{V} + (5\,\text{mA})\,(R_2) \tag{14-9b}$$

For example, if you need a 5 V supply for TTL logic, make $R_2 = 760\ \Omega$. If you need a +15-V supply for an op amp or CMOS, make $R_2 = 2760\ \Omega$. $R_2 = 2160\ \Omega$ will give you 12-V car voltage. Make R_2 a 3-kΩ pot and you can adjust V_o to any voltage between 1.2 V (D-cell battery) and 16.2 V.

14-9.2 Characteristics of the LM317HV

The LM317HV will provide a regulated output current of up to 1.5 A, provided that it is not subjected to a power dissipation of more than about 15 W (TO-3 case). This means it should be electrically isolated from, and fastened to, a large heat sink such as the metal chassis of the power supply. A 5-by 5-in. piece of aluminum chassis stock also makes an adequate heat sink.

The LM317 requires a minimum "dropout" voltage of 3 V across its input and output terminals or it will drop out of regulation. Thus the upper limit of V_o is 3 V below the minimum input voltage from the unregulated supply.

It is good practice to connect bypass capacitors C_1 and C_2 (1-μF tantalum) as shown in Fig. 14-15(a). C_1 minimizes problems caused by long leads between

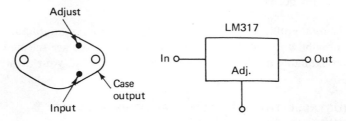

(a) LM317HV connection diagram for TO-3 steel case and circuit schematic

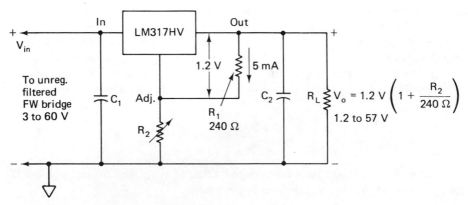

$$V_o = 1.2 \text{ V} \left(1 + \frac{R_2}{240 \ \Omega} \right)$$

(b) Connecting the LM317HV to act as an adjustable positive voltage regulator

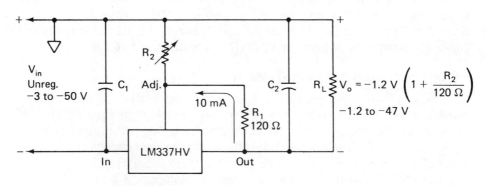

$$V_o = -1.2 \text{ V} \left(1 + \frac{R_2}{120 \ \Omega} \right)$$

(c) Connecting the LM337HV to act as an adjustable negative regulator

Figure 14-15 Adjustable three-terminal positive (LM317) and negative (LM337) IC regulators are easy to use.

the rectifier and LM317. C_2 improves transient response. Any ripple voltage from the rectifier will be reduced by a factor of over 1000 if R_2 is bypassed by a 1-μF tantalum capacitor or 10-μF aluminum electrolytic capacitor.

The LM317HV protects itself against overheating, too much internal power dissipation, and too much current. When the chip temperature reaches 175°C, the 317 shuts down. If the product of output current and input-to-output voltage exceeds 15 to 20 W, or if currents greater than 1.5 A are required, the LM317 also shuts down. When the overload condition is removed, the LM317 simply resumes operation. All of these protection features are made possible by the remarkable internal circuitry of the LM317.

14-9.3 Adjustable Negative-Voltage Regulator

An adjustable three-terminal *negative*-voltage regulator is also available. [see the LM337HV in Fig. 14-15(c)]. It has the pinout shown in Figure 14-15(d) (TO-3 package). The negative regulator operates on the same principle as the positive regulator except that R_1 is a 120-Ω resistor and the maximum input voltage is reduced to 50 V.

V_o is given by Eq. (7-9a). If $R_1 = 120\ \Omega$, then V_o depends upon R_2 according to

$$V_o = 1.2\ \text{V} + (10\ \text{mA})R_2 \tag{14-10}$$

14-10 BIPOLAR REGULATED POWER SUPPLY

In the op amp circuits studied this far, a ±15-V bipolar regulated voltage power supply was necessary. One excellent IC voltage regulator made exactly for this application is the RC 4195 (available from Raytheon Semi-Conductor Division, 350 Ellis Street, Mountain View, California 94040).

To use the 4195, simply connect its input terminals $+V_{\text{in}}$ and $-V_{\text{in}}$ and ground (gnd) to an unregulated supply of ±18 to ±30 V. For example, let the transformer of Fig. 14-7 be a 110 V : 30 V CT, which gives respective unregulated dc voltages of $+21$ V and -21 V at terminals 1 and 2. Make the following connections:

Fig. 14-7	Fig. 14-16
From terminal 1	To $+V_{\text{in}}$
From terminal CT	To gnd
From terminal 2	To $-V_{\text{in}}$

Terminals $+V_o$ and $-V_o$ of the 4195 give $+15$ V and -15 V with respect to ground. The manufacturer recommends installation of two 10-μF bypass capacitors at the output terminals, as shown in Fig. 14-16.

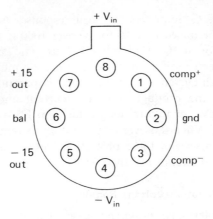

(a) RC4195 Dual regulator TO-99 package, bottom view

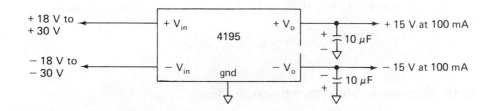

(b) Connections for a ± 15-V voltage regulator

Figure 14-16   The RC 4195, a ±15-V dual-tracking voltage regulator.

The 4195 features thermal shutdown and internal short-circuit protection. A clip-on heat sink should be installed to achieve high load-current capability. A single 4195 IC regulator can supply power to more than 10 op amps.

14-11 POWER AMPLIFIER

The ideas of current boost, the basic op amp regulator, and an inverting amplifier can be put together to form an inexpensive but excellent power amplifier, as shown in Fig. 14-17. The 301 op amp has feed-forward compensation (see Chapter 10) to improve high-frequency response. R_f and R_i set the amplifier's voltage gain at 27. When E_i is negative, Q_1 conducts to furnish a positive output voltage. If the current gain of Q_1 is high ($\beta \geq 100$), then Q_1 boosts the maximum output current capability of the op amp (approximately 10 mA) to over 1 A. When E_i is positive, Q_2 conducts to furnish the output voltage to load R_L.

The frequency response of the amplifier is essentially flat from dc to over

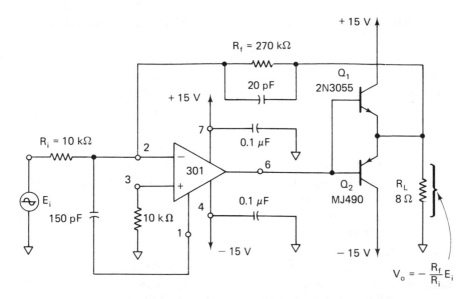

Figure 14-17 A 5-W audio power amplifier.

20 kHz. The power output is a maximum of about 5 W to an 8-Ω or 4-Ω speaker.

The amplifier circuit may also be used for a positive- or negative-voltage regulator for any load voltage between -12 and $+12$ V. This is done by replacing E_i with a dc voltage. The circuit of Fig. 14-17 is connected as an inverting amplifier; therefore, E_i must be opposite in polarity to the desired regulated load voltage. Choose R_f and R_i to give the correct value of V_o in accordance with $V_o = -E_i R_f / R_i$.

PROBLEMS

14-1. A transformer is rated at 115 to 28 V rms at 1 A. What is the peak secondary voltage?

14-2. A 115- to 28-V transformer is used in Fig. 14-4(a). Find $V_{\text{dc no load}}$.

14-3. As dc load current decreases, what happens to (a) dc load voltage; (b) ac ripple voltage?

14-4. In Fig. 14-4, the transformer rating is 115 to 28 V at 1 A. What is V_{dc} at a full load current of $I_L = 0.5$ A?

14-5. Recalculate ΔV in Example 14-3 for a load current of 0.5 A.

14-6. What voltage readings would be obtained with an ac voltmeter for peak-to-peak ripple voltages of (a) 1 V; (b) 3 V?

14-7. The dc full-load voltage of a power supply is 28 V and the peak-to-peak ripple voltage is 6 V. Find the minimum instantaneous load voltage.

14-8. A 110 V: 28 V CT transformer is installed in Fig. 14-7. What no-load dc voltage would be measured (a) between terminals 1 and 2; (b) from 1 to CT; (c) from 2 to CT?

14-9. $R_f = R_i = 10 \text{ k}\Omega$ in Fig. 14-12. Show that $V_o = 10 \text{ V}$.

14-10. In Fig. 14-12, find R_f and R_i for $V_o = 10 \text{ V}$ if $V_z = 5 \text{ V}$.

14-11. Evaluate R_D in Fig. 14-12 if $V_z = 5 \text{ V}$ and $V_o = 10 \text{ V}$.

14-12. If V_o is reduced to 10 V in Example 14-5, find the power dissipation in the pass transistor.

14-13. In Fig. 14-14, $R_S = 1 \Omega$. Find the maximum load current.

14-14. What is the amplifier gain in Fig. 14-17 if R_f is changed to 220 kΩ?

14-15. You need a regulated output voltage of 24.0 V. If $R_1 = 240 \Omega$ in Fig. 14-5(b), find the required value of R_2.

14-16. Assume that the regulator of Problem 14-15 delivers a load current of 1.0 A. If its overage dc input voltage is 30 V, show that the regulator must dissipate 6 W.

14-17. Find V_o if R_2 is short-circuited in (a) Fig. 14-15(b); (b) Fig. 14-15(c).

14-18. Adjustable resistor $R_2 = 0$ to 2500 Ω in Fig. 14-15(b). Find the upper and lower limits of V_o as R_2 is adjusted from 2500 Ω to 0 Ω.

14-19. Suppose that a 1200-Ω low-stop resistor, and a 2500-Ω pot are connected in series in place of the single resistor R_2 in Fig. 14-15(b). Find the upper and lower limits of V_o as the pot is adjusted from 2500 Ω to 0 Ω.

14-20. If the dropout voltage of an LM317 is 3 V, what is the minimum instantaneous input voltage for the regulator circuit of Problem 14-18?

bibliography

ılıılıılıılıılıılıılıılıılıılıılıılıılıılıılıılıılıılıılıılıılıılıılıılıılıılıılıılıılıılıılıılıılıılıılı

ANALOG DEVICES, INC., *Analog-Digital Conversion Handbook* (1972), *Analog-Digital Conversion Notes* (1977), *Data Acquisition Products Catalog* (1978), *Non-linear Circuits Handbook* (1979), *Data Acquisition Products Catalog Supplement* (1979), Analog, Norwood, Mass.

BLH ELECTRONICS, *Strain Gages, SR4*, BLH, Waltham, Mass (1979).

CLAYTON, G. G., *Operational Amplifiers*, Butterworth & Co. (Publishers) Ltd., London (1971).

COUGHLIN, ROBERT F., *Principles and Applications of Semiconductors and Circuits*, Prentice-Hall, Inc., Englewood Cliffs, N.J. (1971).

COUGHLIN, ROBERT F., and DRISCOLL, FREDERICK F., *Semiconductor Fundamentals*, Prentice-Hall, Inc., Englewood Cliffs, N.J. (1976).

DRISCOLL, FREDERICK F., *Analysis of Electric Circuits*, Prentice-Hall, Inc., Englewood Cliffs, N.J. (1973).

DRISCOLL, FREDERICK F., and COUGHLIN, ROBERT F., *Solid State Devices and Applications*, Prentice-Hall, Inc., Englewood Cliffs, N.J. (1975).

FAIRCHILD INSTRUMENT AND CAMERA CORPORATION, *Linear Interface Book* (1978), *Voltage Regulator Handbook* (1974), Fairchild, Mountain View, Calif.

HARRIS CORPORATION, *Linear and Data Acquisition Products*, Harris, Melbourne, Fla. (1977).

MORRISON, R., *Grounding and Shielding Techniques in Instrumentation*, John Wiley & Sons, Inc., New York (1967).

MOTOROLA SEMICONDUCTOR PRODUCTS, INC., *More Value Out of Integrated Operational Amplifier Data Sheets*, AN-273A, Motorola, Phoenix, Ariz. (1970).

NATIONAL SEMICONDUCTOR CORPORATION, *Data Acquisition Handbook* (1978), *Linear Data Book* (1980), *Linear Data Acquisition Handbook* (1980), *Linear Applications Handbook* (1980), *Special Functions Data Book* (1979), *Voltage Regulator Handbook* (1980), *Audio Handbook* (no date), National Semiconductor, Santa Clara, Calif.

PHILBRICK RESEARCHES, INC., *Applications Manual for Computing Amplifiers for Modeling, Measuring, Manipulating and Much Else*, Nimrod Press, Inc., Boston (1966).

PRECISION MONOLITHICS, INC., *Linear and Conversion IC Products*, Precision Monolithics, Santa Clara, Calif. (1978).

SIGNETICS CORPORATION, *Data Manual*, Signetics, Sunnyvale, Calif. (1977).

SMITH, J. I., *Modern Operational Circuit Design*, John Wiley & Sons, Inc., New York (1971).

VILLANUCCI, R., ET AL., *Electronic Techniques: Shop Practices and Construction*, 2nd ed., Prentice-Hall, Inc., Englewood Cliffs, N.J. (1981).

μA 74I
frequency compensated
operational amplifier*

ʋʍ

appendix 1

*Courtesy of Fairchild Semiconductor, a Division of Fairchild Camera and Instrument Corporation.

GENERAL DESCRIPTION — The μA741 is a high performance monolithic Operational Amplifier constructed using the Fairchild Planar* epitaxial process. It is intended for a wide range of analog applications. High common mode voltage range and absence of "latch-up" tendencies make the μA741 ideal for use as a voltage follower. The high gain and wide range of operating voltage provides superior performance in integrator, summing amplifier, and general feedback applications.

- **NO FREQUENCY COMPENSATION REQUIRED**
- **SHORT CIRCUIT PROTECTION**
- **OFFSET VOLTAGE NULL CAPABILITY**
- **LARGE COMMON-MODE AND DIFFERENTIAL VOLTAGE RANGES**
- **LOW POWER CONSUMPTION**
- **NO LATCH UP**

ABSOLUTE MAXIMUM RATINGS

Supply Voltage	
Military (741)	±22 V
Commercial (741C)	±18 V
Internal Power Dissipation (Note 1)	
Metal Can	500 mW
DIP	670 mW
Mini DIP	310 mW
Flatpak	570 mW
Differential Input Voltage	±30 V
Input Voltage (Note 2)	±15 V
Storage Temperature Range	
Metal Can, DIP, and Flatpak	−65°C to +150°C
Mini DIP	−55°C to +125°C
Operating Temperature Range	
Military (741)	−55°C to +125°C
Commercial (741C)	0°C to +70°C
Lead Temperature (Soldering)	
Metal Can, DIP, and Flatpak (60 seconds)	300°C
Mini DIP (10 seconds)	260°C
Output Short Circuit Duration (Note 3)	Indefinite

EQUIVALENT CIRCUIT

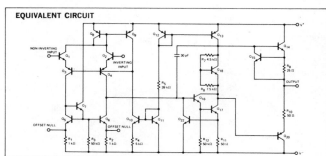

Notes on following pages.

ELECTRICAL CHARACTERISTICS ($V_S = \pm15$ V, $T_A = 25°C$ unless otherwise specified)

PARAMETERS (see definitions)		CONDITIONS	MIN.	TYP.	MAX.	UNITS
Input Offset Voltage		$R_S \leqslant 10 \text{ k}\Omega$		1.0	5.0	mV
Input Offset Current				20	200	nA
Input Bias Current				80	500	nA
Input Resistance			0.3	2.0		$M\Omega$
Input Capacitance				1.4		pF
Offset Voltage Adjustment Range				±15		mV
Large Signal Voltage Gain		$R_L \geqslant 2 \text{ k}\Omega$, $V_{OUT} = \pm10$ V	50,000	200,000		
Output Resistance				75		Ω
Output Short Circuit Current				25		mA
Supply Current				1.7	2.8	mA
Power Consumption				50	85	mW
Transient Response (Unity Gain)	Risetime	$V_{IN} = 20$ mV, $R_L = 2 \text{ k}\Omega$, $C_L \leqslant 100$ pF		0.3		μs
	Overshoot			5.0		%
Slew Rate		$R_L \geqslant 2 \text{ k}\Omega$		0.5		$V/\mu s$

The following specifications apply for $-55°C \leqslant T_A \leqslant +125°C$:

Input Offset Voltage		$R_S \leqslant 10 \text{ k}\Omega$		1.0	6.0	mV
Input Offset Current	$T_A = +125°C$			7.0	200	nA
	$T_A = -55°C$			85	500	nA
Input Bias Current	$T_A = +125°C$			0.03	0.5	μA
	$T_A = -55°C$			0.3	1.5	μA
Input Voltage Range			±12	±13		V
Common Mode Rejection Ratio		$R_S \leqslant 10 \text{ k}\Omega$	70	90		dB
Supply Voltage Rejection Ratio		$R_S \leqslant 10 \text{ k}\Omega$		30	150	$\mu V/V$
Large Signal Voltage Gain		$R_L \geqslant 2 \text{ k}\Omega$, $V_{OUT} = \pm10$ V	25,000			
Output Voltage Swing	$R_L \geqslant 10 \text{ k}\Omega$		±12	±14		V
	$R_L \geqslant 2 \text{ k}\Omega$		±10	±13		V
Supply Current	$T_A = +125°C$			1.5	2.5	mA
	$T_A = -55°C$			2.0	3.3	mA
Power Consumption	$T_A = +125°C$			45	75	mW
	$T_A = -55°C$			60	100	mW

TYPICAL PERFORMANCE CURVES FOR 741

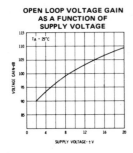

OPEN LOOP VOLTAGE GAIN AS A FUNCTION OF SUPPLY VOLTAGE

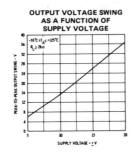

OUTPUT VOLTAGE SWING AS A FUNCTION OF SUPPLY VOLTAGE

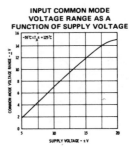

INPUT COMMON MODE VOLTAGE RANGE AS A FUNCTION OF SUPPLY VOLTAGE

ELECTRICAL CHARACTERISTICS (V_S = ±15 V, T_A = 25°C unless otherwise specified)

PARAMETERS (see definitions)		CONDITIONS	MIN.	TYP.	MAX.	UNITS
Input Offset Voltage		$R_S \leqslant 10\ k\Omega$		2.0	6.0	mV
Input Offset Current				20	200	nA
Input Bias Current				80	500	nA
Input Resistance			0.3	2.0		MΩ
Input Capacitance				1.4		pF
Offset Voltage Adjustment Range				±15		mV
Input Voltage Range			±12	±13		V
Common Mode Rejection Ratio		$R_S \leqslant 10\ k\Omega$	70	90		dB
Supply Voltage Rejection Ratio		$R_S \leqslant 10\ k\Omega$		30	150	μV/V
Large Signal Voltage Gain		$R_L \geqslant 2\ k\Omega, V_{OUT} = \pm10\ V$	20,000	200,000		
Output Voltage Swing		$R_L \geqslant 10\ k\Omega$	±12	±14		V
		$R_L \geqslant 2\ k\Omega$	±10	±13		V
Output Resistance				75		Ω
Output Short Circuit Current				25		mA
Supply Current				1.7	2.8	mA
Power Consumption				50	85	mW
Transient Response (Unity Gain)	Risetime	V_{IN} = 20 mV, R_L = 2 kΩ, $C_L \leqslant$ 100 pF		0.3		μs
	Overshoot			5.0		%
Slew Rate		$R_L \geqslant 2\ k\Omega$		0.5		V/μs

The following specifications apply for 0°C $\leqslant T_A \leqslant$ +70°C:

Input Offset Voltage					7.5	mV
Input Offset Current					300	nA
Input Bias Current					800	nA
Large Signal Voltage Gain		$R_L \geqslant 2\ k\Omega, V_{OUT} = \pm10\ V$	15,000			
Output Voltage Swing		$R_L \geqslant 2\ k\Omega$	±10	±13		V

TYPICAL PERFORMANCE CURVES FOR 741C

OPEN LOOP VOLTAGE GAIN AS A FUNCTION OF SUPPLY VOLTAGE

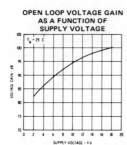

OUTPUT VOLTAGE SWING AS A FUNCTION OF SUPPLY VOLTAGE

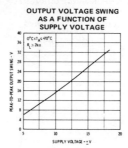

INPUT COMMON MODE VOLTAGE RANGE AS A FUNCTION OF SUPPLY VOLTAGE

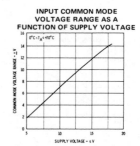

NOTES:
1. Rating applies to ambient temperatures up to 70°C. Above 70°C ambient derate linearly at 6.3 mW/°C for the Metal Can, 8.3 mW/°C for the DIP, 5.6 mW/°C for the Mini DIP and 7.1 mW/°C for the Flatpak.
2. For supply voltages less than ±15 V, the absolute maximum input voltage is equal to the supply voltage.
3. Short circuit may be to ground or either supply. Rating applies to +125°C case temperature or 75°C ambient temperature.

TYPICAL PERFORMANCE CURVES FOR 741

INPUT BIAS CURRENT AS A FUNCTION OF AMBIENT TEMPERATURE

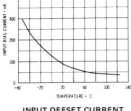

INPUT RESISTANCE AS A FUNCTION OF AMBIENT TEMPERATURE

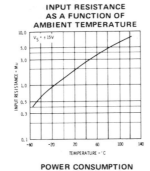

OUTPUT SHORT-CIRCUIT CURRENT AS A FUNCTION OF AMBIENT TEMPERATURE

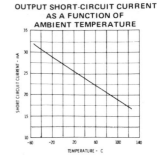

INPUT OFFSET CURRENT AS A FUNCTION OF AMBIENT TEMPERATURE

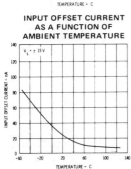

POWER CONSUMPTION AS A FUNCTION OF AMBIENT TEMPERATURE

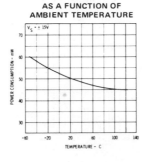

FREQUENCY CHARACTERISTICS AS A FUNCTION OF AMBIENT TEMPERATURE

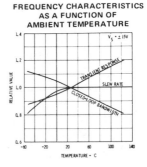

TYPICAL PERFORMANCE CURVES FOR 741C

INPUT BIAS CURRENT AS A FUNCTION OF AMBIENT TEMPERATURE

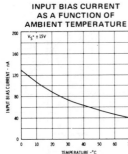

INPUT RESISTANCE AS A FUNCTION OF AMBIENT TEMPERATURE

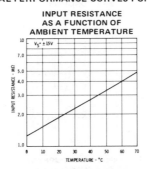

INPUT OFFSET CURRENT AS A FUNCTION OF AMBIENT TEMPERATURE

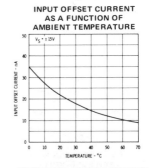

POWER CONSUMPTION AS A FUNCTION OF AMBIENT TEMPERATURE

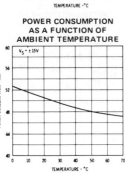

OUTPUT SHORT-CIRCUIT CURRENT AS A FUNCTION OF AMBIENT TEMPERATURE

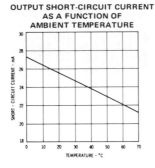

FREQUENCY CHARACTERISTICS AS A FUNCTION OF AMBIENT TEMPERATURE

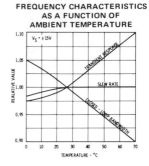

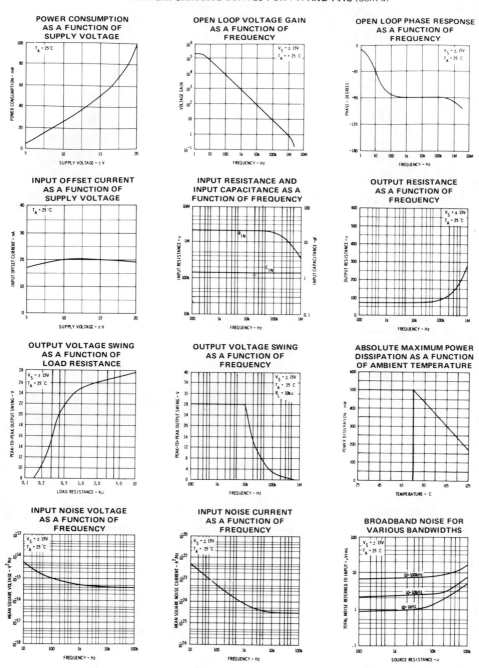

TRANSIENT RESPONSE

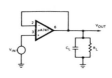

OUTPUT – mV

90%

10%
RISE TIME

$V_S = \pm 15V$
$T_A = 25°C$
$R_L = 2k\Omega$
$C_L = 100pF$

TIME – µs

TRANSIENT RESPONSE TEST CIRCUIT

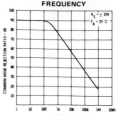

V_{OUT}

V_{IN}

COMMON MODE REJECTION RATIO AS A FUNCTION OF FREQUENCY

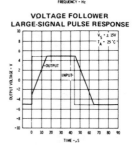

COMMON MODE REJECTION RATIO – dB

$V_S = \pm 15V$
$T_A = 25°C$

FREQUENCY – Hz

FREQUENCY CHARACTERISTICS AS A FUNCTION OF SUPPLY VOLTAGE

RELATIVE VALUE

$T_A = 25°C$

TRANSIENT RESPONSE

SLEW RATE

CLOSED LOOP BANDWIDTH

SUPPLY VOLTAGE – ± V

VOLTAGE OFFSET NULL CIRCUIT

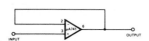

µA741

10 kΩ

V^-

VOLTAGE FOLLOWER LARGE-SIGNAL PULSE RESPONSE

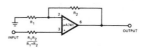

OUTPUT VOLTAGE – V

$V_S = \pm 15V$
$T_A = 25°C$

OUTPUT

INPUT

TIME – µS

TYPICAL APPLICATIONS

UNITY-GAIN VOLTAGE FOLLOWER

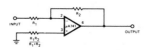

INPUT   µA741   OUTPUT

$R_{IN} = 400\ M\Omega$
$C_{IN} = 1\ pF$
$R_{OUT} << 1\ \Omega$
B.W. = 1 MHz

NON-INVERTING AMPLIFIER

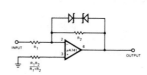

INPUT   µA741   OUTPUT

$\frac{R_1 R_2}{R_1 + R_2}$

GAIN	R_1	R_2	B.W.	R_{IN}
10	1 kΩ	9 kΩ	100 kHz	400 MΩ
100	100 Ω	9.9 kΩ	10 kHz	280 MΩ
1000	100 Ω	99.9 kΩ	1 kHz	80 MΩ

INVERTING AMPLIFIER

INPUT R_1 µA741   OUTPUT R_2

$\frac{R_1 R_2}{R_1 + R_2}$

GAIN	R_1	R_2	B.W.	R_{IN}
1	10 kΩ	10 kΩ	1 MHz	10 kΩ
10	1 kΩ	10 kΩ	100 kHz	1 kΩ
100	1 kΩ	100 kΩ	10 kHz	1 kΩ
1000	100 Ω	100 kΩ	1 kHz	100 Ω

CLIPPING AMPLIFIER

INPUT R_1 µA741   OUTPUT R_2

$\frac{R_1 R_2}{R_1 + R_2}$

$$\frac{E_{OUT}}{E_{IN}} = \frac{R_2}{R_1} \text{ if } |E_{OUT}| \leq V_Z + 0.7\ V$$

where V_Z = Zener breakdown voltage

SIMPLE INTEGRATOR

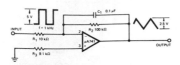

$$E_{OUT} = -\frac{1}{R_1 C_1}\int E_{IN}dt$$

SIMPLE DIFFERENTIATOR

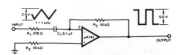

$$E_{OUT} = -R_2 C_1 \frac{dE_{IN}}{dt}$$

LOW DRIFT LOW NOISE AMPLIFIER

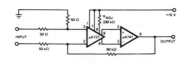

Voltage Gain = 10^3
Input Offset Voltage Drift = 0.6 $\mu V/°C$
Input Offset Current Drift = 2.0 $pA/°C$

HIGH SLEW RATE POWER AMPLIFIER

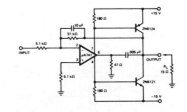

NOTCH FILTER USING THE µA741 AS A GYRATOR

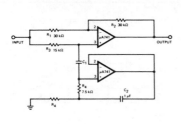

Trim R_3 such that
$$\frac{R_1}{R_2} = \frac{R_3}{2 R_4}$$

NOTCH FREQUENCY AS A FUNCTION OF C_1

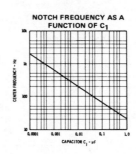

PHYSICAL DIMENSIONS

(H) 5B 8-LEAD METAL CAN
in accordance with JEDEC (TO-99) outline

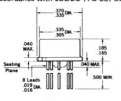

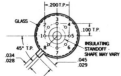

NOTES:
All dimensions in inches
Leads are gold-plated kovar
Package weight is 1.22 gram

(D) 6A 14-LEAD HERMETIC DIP
in accordance with JEDEC (TO-116) outline

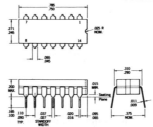

NOTES:
All dimensions in inches
Leads are intended for insertion in hole rows on .300" centers
They are purposely shipped with "positive" misalignment to facilitate insertion
Board-drilling dimensions should equal your practice for .020 inch diameter lead
Leads are tin-plated kovar
Package weight is 2.0 grams

(T) 9T 8-LEAD MINI DIP

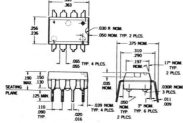

NOTES:
All dimensions in inches
Leads are intended for insertion in hole rows on .300" centers
They are purposely shipped with "positive" misalignment to facilitate insertion
Board-drilling dimensions should equal your practice for .020 inch diameter lead
Leads are tin or gold-plated kovar
Package weight is 0.6 gram

(F) 3F 10-LEAD FLATPAK
in accordance with JEDEC (TO-91) outline

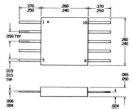

NOTES:
All dimensions in inches
Leads are gold-plated kovar
Package weight is 0.26 gram

LM301A operational amplifier*

appendix 2

*Courtesy of National Semiconductor Corporation.

general description

The LM301A is a general-purpose operational amplifier which features improved performance over the 709C and other popular amplifiers. Advanced processing techniques make possible an order of magnitude reduction in input currents, and a redesign of the biasing circuitry reduces the temperature drift of input current.

This amplifier offers many features which make its application nearly foolproof: overload protection on the input and output, no latch-up when the common mode range is exceeded, freedom from oscillations and compensation with a single 30 pF capacitor. It has advantages over internally compensated amplifiers in that the compensation can be tailored to the particular application. For example, as a summing amplifier, slew rates of 10 V/μs and bandwidths of 10 MHz can be realized. In addition, the circuit can be used as a comparator with differential inputs up to ±30V; and the output can be clamped at any desired level to make it compatible with logic circuits.

The LM301A provides better accuracy and lower noise than its predecessors in high impedance circuitry. The low input currents also make it particularly well suited for long interval integrators or timers, sample and hold circuits and low frequency waveform generators. Further, replacing circuits where matched transistor pairs buffer the inputs of conventional IC op amps, it can give lower offset voltage and drift at reduced cost.

typical applications

Fast Summing Amplifier

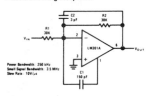

Fast Voltage Follower

Standard Compensation and Offset Balancing Circuit

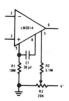

Integrator with Bias Current Compensation

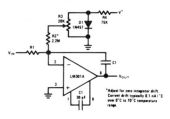

Low Frequency Square Wave Generator

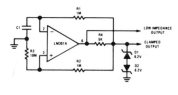

Bilateral Current Source

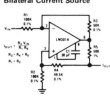

Voltage Comparator for Driving DTL or TTL Integrated Circuits

Double-Ended Limit Detector

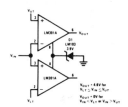

absolute maximum ratings

Supply Voltage	±18V
Power Dissipation (Note 1)	500 mW
Differential Input Voltage	±30V
Input Voltage (Note 2)	±15V
Output Short-Circuit Duration (Note 3)	Indefinite
Operating Temperature Range	0°C to 70°C
Storage Temperature Range	−65°C to 150°C
Lead Temperature (Soldering, 60 sec)	300°C

electrical characteristics (Note 4)

PARAMETER	CONDITIONS	MIN	TYP	MAX	UNITS
Input Offset Voltage	T_A = 25°C, $R_S \leq$ 50 kΩ		2.0	7.5	mV
Input Offset Current	T_A = 25°C		3	50	nA
Input Bias Current	T_A = 25°C		70	250	nA
Input Resistance	T_A = 25°C	0.5	2		MΩ
Supply Current	T_A = 25°C, V_S = ±15V		1.8	3.0	mA
Large Signal Voltage Gain	T_A = 25°C, V_S = ±15V V_{OUT} = ±10V, $R_L \geq$ 2 kΩ	25	160		V/mV
Input Offset Voltage	$R_S \leq$ 50 kΩ			10	mV
Average Temperature Coefficient of Input Offset Voltage			6.0	30	µV/°C
Input Offset Current				70	nA
Average Temperature Coefficient of Input Offset Current	25°C $\leq T_A \leq$ 70°C 0°C $\leq T_A \leq$ 25°C		0.01 0.02	0.3 0.6	nA/°C nA/°C
Input Bias Current				300	nA
Large Signal Voltage Gain	V_S = ±15V, V_{OUT} = ±10V $R_L \geq$ 2 kΩ	15			V/mV
Output Voltage Swing	V_S = ±15V, R_L = 10 kΩ R_L = 2 kΩ	±12 ±10	±14 ±13		V V
Input Voltage Range	V_S = ±15V	±12			V
Common Mode Rejection Ratio	$R_S \leq$ 50 kΩ	70	90		dB
Supply Voltage Rejection Ratio	$R_S \leq$ 50 kΩ	70	96		dB

Note 1: For operating at elevated temperatures, the device must be derated based on a 100°C maximum junction temperature and a thermal resistance of 150°C/W junction to ambient or 45°C/W junction to case.

Note 2: For supply voltages less than ±15V, the absolute maximum input voltage is equal to the supply voltage.

Note 3: Continuous short circuit is allowed for case temperatures to 70°C and ambient temperatures to 55°C.

Note 4: These specifications apply for 0°C $\leq T_A <$ 70°C, ±5V, $\leq V_S \leq$ ±15V and C1 = 30 pF unless otherwise specified.

guaranteed performance

Input Voltage Range

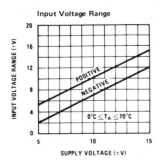

Output Swing

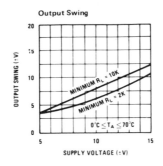

Voltage Gain

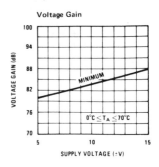

typical performance

Supply Current

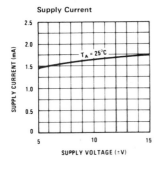

Voltage Gain

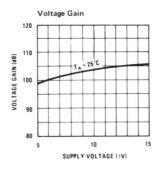

Input Current

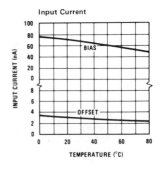

Current Limiting

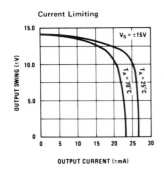

Input Noise Voltage

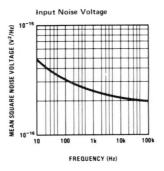

Input Noise Current

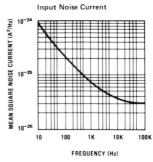

Open Loop Frequency Response

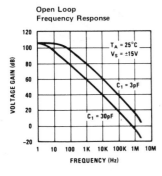

Large Signal Frequency Response

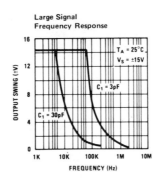

Voltage Follower Pulse Response

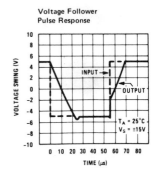

definition of terms

Input Offset Voltage: That voltage which must be applied between the input terminals through two equal resistances to obtain zero output voltage.

Input Offset Current: The difference in the currents into the two input terminals when the output is at zero.

Input Voltage Range: The range of voltages on the input terminals for which the offset specifications apply.

Input Bias Current: The average of the two input currents.

Common Mode Rejection Ratio: The ratio of the input voltage range to the peak-to-peak change in input offset voltage over this range.

Input Resistance: The ratio of the change in input voltage to the change in input current on either input with the other grounded.

Supply Current: The current required from the power supply to operate the amplifier with no load and the output at zero.

Output Voltage Swing: The peak output voltage swing, referred to zero, that can be obtained without clipping.

Large-Signal Voltage Gain: The ratio of the output voltage swing to the change in input voltage required to drive the output from zero to this voltage.

Power Supply Rejection: The ratio of the change in input offset voltage to the change in power supply voltages producing it.

physical dimensions

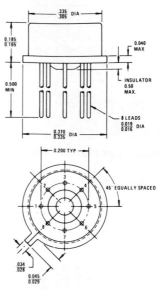

Order Number LM301AH

connection diagram

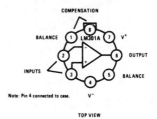

Note: Pin 4 connected to case.

TOP VIEW

LM311 voltage comparator*

ⲓⲁⲓⲩⲓⲁⲓⲩⲓⲁⲓⲩⲓⲁⲓⲩⲓⲁⲓⲩⲓⲁⲓⲩⲓⲁⲓⲩⲓⲁⲓⲩⲓⲁⲓⲩⲓⲁⲓⲩⲓⲁⲓⲩⲓⲁⲓⲩⲓⲁⲓⲩⲓⲁⲓⲩⲓⲁⲓⲩⲓⲁⲓⲩ

appendix 3

*Courtesy of National Semiconductor Corporation.

general description

The LM311 is a voltage comparator that has input currents more than a hundred times lower than devices like the LM306 or LM710C. It is also designed to operate over a wider range of supply voltages: from standard ±15V op amp supplies down to the single 5V supply used for IC logic. Its output is compatible with RTL, DTL and TTL as well as MOS circuits. Further, it can drive lamps or relays, switching voltages up to 40V at currents as high as 50 mA. Outstanding characteristics include:

- Maximum input current: 250 nA

- Maximum offset current: 50 nA

- Differential input voltage range: ±30V

- Power consumption: 135 mW at ±15V

Both the input and the output of the LM311 can be isolated from system ground, and the output can drive loads referred to ground, the positive supply or the negative supply. Offset balancing and strobe capability are provided and outputs can be wire OR'ed. Although slower than the LM306 and LM710C (200 ns response time vs 40 ns) the device is also much less prone to spurious oscillations. The LM311 has the same pin configuration as the LM306 and LM710C.

schematic diagram and auxiliary circuits

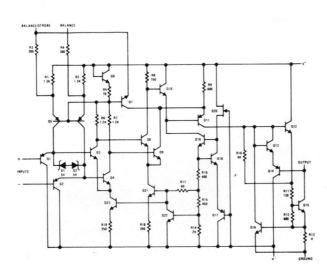

Offset Balancing

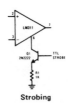

Strobing

Increasing Input Stage Current*

typical applications

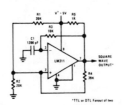

100 kHz Free Running Multivibrator

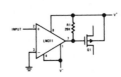

Zero Crossing Detector Driving MOS Switch

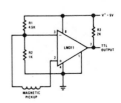

Detector for Magnetic Transducer

344

absolute maximum ratings

Total Supply Voltage (V_{84})	36V
Output to Negative Supply Voltage (V_{74})	40V
Ground to Negative Supply Voltage (V_{14})	30V
Differential Input Voltage	±30V
Input Voltage (Note 1)	±15V
Power Dissipation (Note 2)	500 mW
Output Short Circuit Duration	10 sec
Operating Temperature Range	0°C to 70°C
Storage Temperature Range	−65°C to 150°C
Lead Temperature (soldering, 10 sec)	300°C

electrical characteristics (Note 3)

PARAMETER	CONDITIONS	MIN	TYP	MAX	UNITS
Input Offset Voltage (Note 4)	$T_A = 25°C$		2.0	7.5	mV
Input Offset Current (Note 4)	$T_A = 25°C$		6.0	50	nA
Input Bias Current	$T_A = 25°C$		100	250	nA
Voltage Gain	$T_A = 25°C$		200		V/mV
Response Time (Note 5)	$T_A = 25°C$		200		ns
Saturation Voltage	$V_{IN} \leq -10$ mV, $I_{OUT} = 50$ mA $T_A = 25°C$		0.75	1.5	V
Output Leakage Current	$V_{IN} \geq 10$ mV, $V_{OUT} = 35$V $T_A = 25°C$		0.2	50	nA
Input Offset Voltage (Note 4)				10	mV
Input Offset Current (Note 4)				70	nA
Input Bias Current				300	nA
Input Voltage Range		±14			V
Saturation Voltage	$V_{IN} \leq -10$ mV, $I_{SINK} \leq 8$ mA		0.23	0.4	V
Positive Supply Current	$T_A = 25°C$		5.1	7.5	mA
Negative Supply Current	$T_A = 25°C$		4.1	5.0	mA

Note 1. This rating applies for ±15V supplies. The positive input voltage limit is 30V above the negative supply. The negative input voltage limit is equal to the negative supply voltage or 30V below the positive supply, whichever is less.

Note 2. The maximum junction temperature of the LM311 is 85°C. For operating at elevated temperatures, devices in the TO-5 package must be derated based on a thermal resistance of 150°C/W, junction to ambient, or 45°C/W, junction to case. For the flat package, the derating is based on a thermal resistance of 185°C/W when mounted on a 1/16-inch-thick epoxy glass board with ten, 0.03-inch-wide, 2-ounce copper conductors. The thermal resistance of the dual-in-line package is 100°C/W, junction to ambient.

Note 3. These specifications apply for $V_S = \pm15V$ and $0°C \leq T_A \leq 70°C$, unless otherwise specified.

Note 4. The offset voltages and offset currents given are the maximum values required to drive the output down to 1V or up to 14V with a 7.5 kΩ load. Thus, these parameters define an error band and take into account the worst case effects of voltage gain and input impedance.

Note 5. The response time specified (see definitions) is for a 100 mV input step with 5 mV overdrive.

typical performance

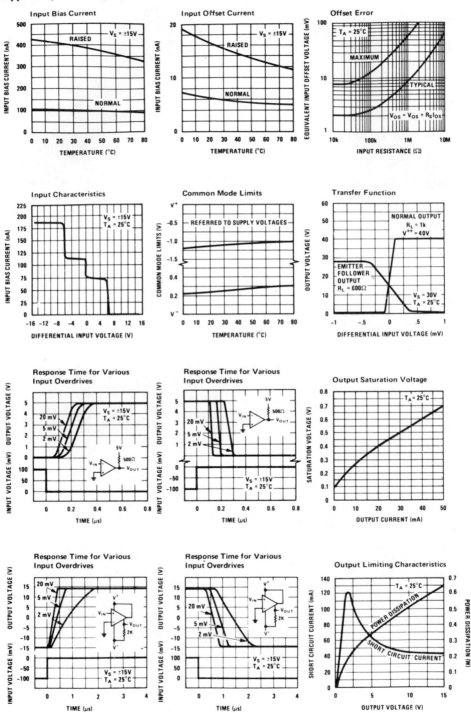

typical performance

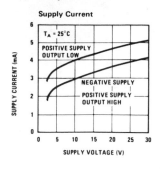

Supply Current

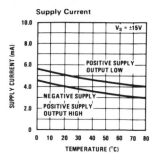

Supply Current

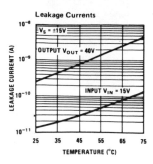

Leakage Currents

typical applications

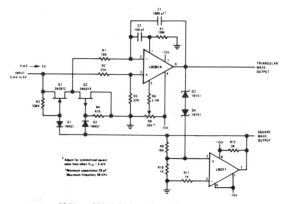

10 Hz to 10 kHz Voltage Controlled Oscillator

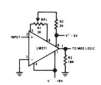

Low Voltage Adjustable Reference Supply

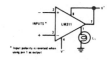

Zero Crossing Detector driving MOS logic

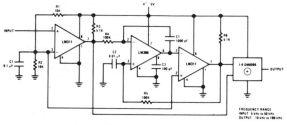

Frequency Doubler

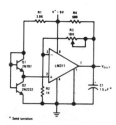

Driving Ground-Referred Load

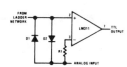

Using Clamp Diodes to Improve Response

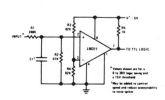

TTL Interface with High Level Logic

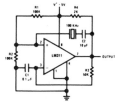

Crystal Oscillator

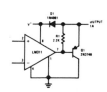

Comparator and Solenoid Driver

typical applications

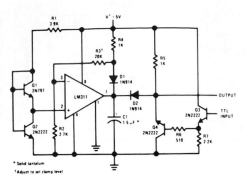

Precision Squarer

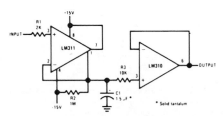

Positive Peak Detector

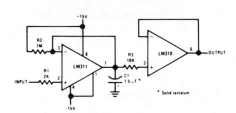

Negative Peak Dectector

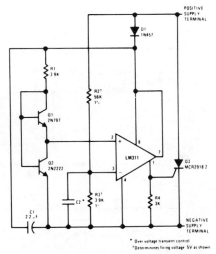

Crowbar Over-Voltage Protector

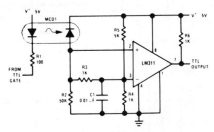

Digital Transmission Isolator

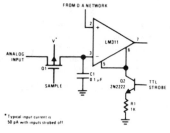

Strobing off Both Input*
and Output Stages

* Typical input current is
50 pA with inputs strobed off

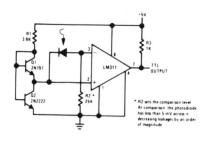

* R2 sets the comparison level
At comparison the photodiode
has less than 5 mV across it
decreasing leakages by an order
of magnitude

Precision Photodiode Comparator

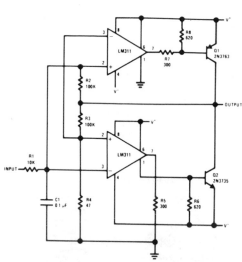

Switching Power Amplifier

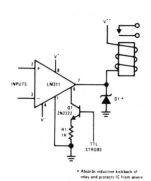

* Absorbs inductive kickback of
relay and protects IC from severe
voltage transients on V⁺⁺ line

Relay Driver with Strobe

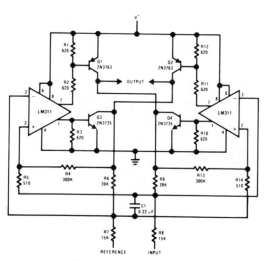

Switching Power Amplifier

definition of terms

Input Offset Voltage: The voltage between the input terminals required to make the output voltage greater than or less than specified voltages.

Input Offset Current: The difference between the two input currents for which the output will be driven higher than or lower than specified voltages.

Input Bias Current: The average of the two input currents.

Input Voltage Range: The range of voltage on the input terminals (common mode) over which all specifications apply.

Voltage Gain: The ratio of the change in output voltage to the change in voltage between the input terminals producing it.

Response Time: The interval between the application of an input step function and the time when the output crosses the logic threshold voltage. The input step drives the comparator from some initial, saturated input voltage to an input level just barely in excess of that required to bring the output from saturation to the logic threshold voltage. This excess is referred to as the voltage overdrive.

Saturation Voltage: The low output voltage level with the input drive equal to or greater than a specified value.

Output Leakage Current: The current into the output terminal with a specified output voltage relative to ground and the input drive equal to or greater than a given value.

Supply Current: The current required from the positive or negative supply to operate the comparator with no output load. The power will vary with input voltage, but is specified as a maximum for the entire range of input voltage conditions.

connection diagrams

Metal Can

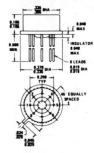

NOTE: Pin 4 connected to case
TOP VIEW

Flat Package

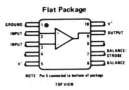

NOTE: Pin 5 connected to bottom of package
TOP VIEW

Dual-In-Line

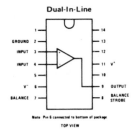

Note: Pin 6 connected to bottom of package
TOP VIEW

physical dimensions

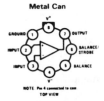

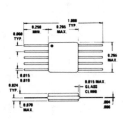

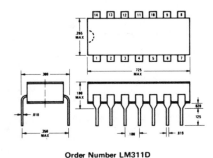

Order Number LM311H   **Order Number LM311F**   **Order Number LM311D**

timer 555*

appendix 4

*Courtesy of Signetics Corporation, 811 East Arques Avenue, Sunnyvale, California, 94086, copyright 1974.

DESCRIPTION

The NE/SE 555 monolithic timing circuit is a highly stable controller capable of producing accurate time delays, or oscillation. Additional terminals are provided for triggering or resetting if desired. In the time delay mode of operation, the time is precisely controlled by one external resistor and capacitor. For a stable operation as an oscillator, the free running frequency and the duty cycle are both accurately controlled with two external resistors and one capacitor. The circuit may be triggered and reset on falling waveforms, and the output structure can source or sink up to 200mA or drive TTL circuits.

FEATURES

- TIMING FROM MICROSECONDS THROUGH HOURS
- OPERATES IN BOTH ASTABLE AND MONOSTABLE MODES
- ADJUSTABLE DUTY CYCLE
- HIGH CURRENT OUTPUT CAN SOURCE OR SINK 200mA
- OUTPUT CAN DRIVE TTL
- TEMPERATURE STABILITY OF 0.005% PER °C
- NORMALLY ON AND NORMALLY OFF OUTPUT

APPLICATIONS

PRECISION TIMING
PULSE GENERATION
SEQUENTIAL TIMING
TIME DELAY GENERATION
PULSE WIDTH MODULATION
PULSE POSITION MODULATION
MISSING PULSE DETECTOR

PIN CONFIGURATIONS (Top View)

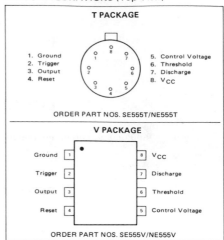

T PACKAGE

1. Ground
2. Trigger
3. Output
4. Reset
5. Control Voltage
6. Threshold
7. Discharge
8. V_{CC}

ORDER PART NOS. SE555T/NE555T

V PACKAGE

Ground	1		8	V_{CC}
Trigger	2		7	Discharge
Output	3		6	Threshold
Reset	4		5	Control Voltage

ORDER PART NOS. SE555V/NE555V

ABSOLUTE MAXIMUM RATINGS

Supply Voltage	+18V
Power Dissipation	600 mW
Operating Temperature Range	
NE555	0°C to +70°C
SE555	−55°C to +125°C
Storage Temperature Range	−65°C to +150°C
Lead Temperature (Soldering, 60 seconds)	+300°C

BLOCK DIAGRAM

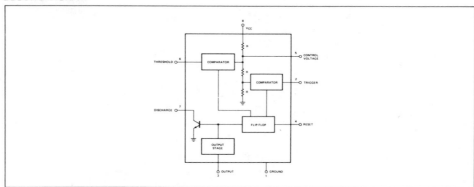

ELECTRICAL CHARACTERISTICS T_A = 25°C, V_{CC} = +5V to +15 unless otherwise specified

PARAMETER	TEST CONDITIONS	SE 555			NE 555			UNITS
		MIN	TYP	MAX	MIN	TYP	MAX	
Supply Voltage		4.5		18	4.5		16	V
Supply Current	V_{CC} = 5V R_L = ∞		3	5		3	6	mA
	V_{CC} = 15V R_L = ∞		10	12		10	15	mA
	Low State, Note 1							
Timing Error(Monostable)	R_A, R_B = 1KΩ to 100KΩ							
Initial Accuracy	C = 0.1 µF Note 2		0.5	2		1		%
Drift with Temperature			30	100		50		ppm/°C
Drift with Supply Voltage			0.05	0.2		0.1		%/Volt
Threshold Voltage			2/3			2/3		X V_{CC}
Trigger Voltage	V_{CC} = 15V	4.8	5	5.2		5		V
Timing Error(Astable)	V_{CC} = 5V	1.45	1.67	1.9		1.67		V
Trigger Current			0.5			0.5		µA
Reset Voltage		0.4	0.7	1.0	0.4	0.7	1.0	V
Reset Current			0.1			0.1		mA
Threshold Current	Note 3		0.1	.25		0.1	.25	µA
Control Voltage Level	V_{CC} = 15V	9.6	10	10.4	9.0	10	11	V
	V_{CC} = 5V	2.9	3.33	3.8	2.6	3.33	4	V
Output Voltage (low)	V_{CC} = 15V							
	I_{SINK} = 10mA		0.1	0.15		0.1	.25	V
	I_{SINK} = 50mA		0.4	0.5		0.4	.75	V
	I_{SINK} = 100mA		2.0	2.2		2.0	2.5	V
	I_{SINK} = 200mA		2.5			2.5		
	V_{CC} = 5V							
	I_{SINK} = 8mA		0.1	0.25				V
	I_{SINK} = 5mA					.25	.35	
Output Voltage Drop (low)								
	I_{SOURCE} = 200mA V_{CC} = 15V		12.5			12.5		
	I_{SOURCE} = 100mA							
	V_{CC} = 15V	13.0	13.3		12.75	13.3		V
	V_{CC} = 5V	3.0	3.3		2.75	3.3		V
Rise Time of Output			100			100		nsec
Fall Time of Output			100			100		nsec

NOTES

1. Supply Current when output high typically 1mA less.

2. Tested at V_{CC} = 5V and V_{CC} = 15V

3. This will determine the maximum value of R_A + R_B. For 15V operation, the max total R = 20 megohm.

EQUIVALENT CIRCUIT (Shown for One Side Only)

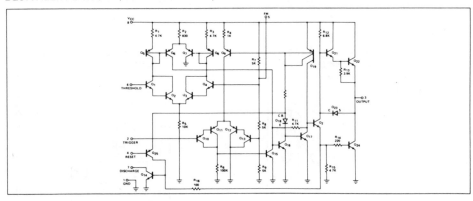

TYPICAL CHARACTERISTICS

MINIMUM PULSE WIDTH REQUIRED FOR TRIGGERING

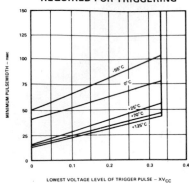

SUPPLY CURRENT vs SUPPLY VOLTAGE

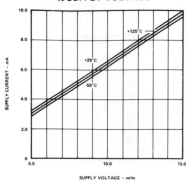

LOW OUTPUT VOLTAGE vs OUTPUT SINK CURRENT

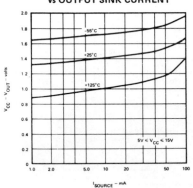

HIGH OUTPUT VOLTAGE vs OUTPUT SOURCE CURRENT

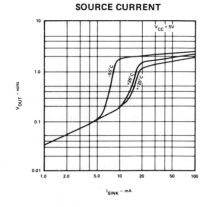

LOW OUTPUT VOLTAGE vs OUTPUT SINK CURRENT

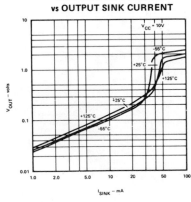

LOW OUTPUT VOLTAGE vs OUTPUT SINK CURRENT

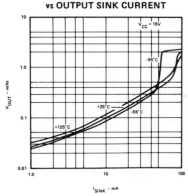

TYPICAL CHARACTERISTICS (Cont'd)

DELAY TIME vs SUPPLY VOLTAGE

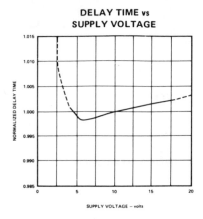

DELAY TIME vs TEMPERATURE

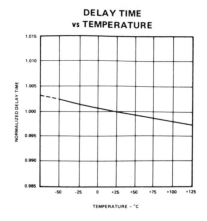

PROPAGATION DELAY vs VOLTAGE LEVEL OF TRIGGER PULSE

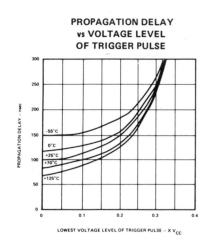

programmable
timer/counter*

appendix 5

*Courtesy of Exar Integrated Systems, Inc.

The XR-2240 Programmable Timer/Counter is a monolithic controller capable of producing ultra-long time delays without sacrificing accuracy. In most applications, it provides a direct replacement for mechanical or electromechanical timing devices and generates programmable time delays from micro-seconds up to five days. Two timing circuits can be cascaded to generate time delays up to three years.

As shown in Figure 1, the circuit is comprised of an internal time-base oscillator, a programmable 8-bit counter and a control flip-flop. The time delay is set by an external R-C network and can be programmed to any value from 1 RC to 255 RC.

In astable operation, the circuit can generate 256 separate frequencies or pulse-patterns from a single RC setting and can be synchronized with external clock signals. Both the control inputs and the outputs are compatible with TTL and DTL logic levels.

FEATURES

Timing from micro-seconds to days
Programmable delays: 1 RC to 255 RC
Wide supply range: 4V to 15V
TTL and DTL compatible outputs
High accuracy: 0.5%
External Sync and Modulation Capability
Excellent Supply Rejection: 0.2%/V

APPLICATIONS

Precision Timing
Long Delay Generation
Sequential Timing
Binary Pattern Generation
Frequency Synthesis
Pulse Counting/Summing
A/D Conversion
Digital Sample and Hold

ABSOLUTE MAXIMUM RATINGS

Supply Voltage	18V
Power Dissipation	
Ceramic Package	750 mW
Derate above +25°C	6 mW/°C
Plastic Package	625 mW
Derate above +25°C	5.0 mW/°C
Operating Temperature	
XR–2240M	−55°C to +125°C
XR–2240C	0°C to +75°C
Storage Temperature	−65°C to +150°C

PACKAGE INFORMATION

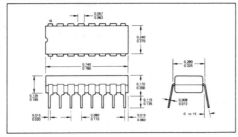

FUNCTIONAL BLOCK DIAGRAM

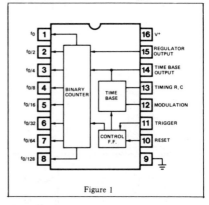

Figure 1

ELECTRICAL CHARACTERISTICS

Test Conditions: See Figure 2, V^+ = 5V, $T_A = 25°C$, R = 10 kΩ, C = 0.1 μF, unless otherwise noted.

PARAMETERS	XR-2240			XR-2240C			UNIT	CONDITIONS
	MIN.	TYP.	MAX.	MIN.	TYP.	MAX.		
GENERAL CHARACTERISTICS								
Supply Voltage	4		15	4		15	V	For V^+ < 4.5V, Short Pin 15 to Pin 16
Supply Current								
Total Circuit		3.5	6		4	7	mA	V^+ = 5V, V_{TR} = 0, V_{RS} = 5V
		12	16		13	18	mA	V^+ = 15V, V_{TR} = 0, V_{RS} = 5V
Counter Only		1			1.5		mA	See Figure 3
Regulator Output, V_R	4.1	4.4		3.9	4.4		V	Measured at Pin 15, V^+ = 5V
	6.0	6.3	6.6	5.8	6.3	6.8	V	V^+ = 15V, See Figure 4
TIME BASE SECTION								See Figure 2
Timing Accuracy *		0.5	2.0		0.5	5	%	V_{RS} = 0, V_{TR} = 5V
Temperature Drift		150	300		200		ppm/°C	V^+ = 5V 0°C ≤ T ≤ 75°C
		80			80		ppm/°C	V^+ = 15V
Supply Drift		0.05	0.2		0.08	0.3	%/V	V^+ ≥ 8 Volts, See Figure 11
Max. Frequency	100	130			130		kHz	R = 1 kΩ, C = 0.007 μF
Modulation Voltage								Measured at Pin 12
Level	3.00	3.50	4.0	2.80	3.50	4.20	V	V^+ = 5V
		10.5			10.5		V	V^+ = 15V
Recommended Range of Timing Components								See Figure 8
Timing Resistor, R	0.001		10	0.001		10	MΩ	
Timing Capacitor, C	0.007		1000	0.01		1000	μF	
TRIGGER/RESET CONTROLS								
Trigger								Measures at Pin 11, V_{RS} = 0
Trigger Threshold		1.4	2.0		1.4	2.0	V	
Trigger Current		8			10		μA	V_{RS} = 0, V_{TR} = 2V
Impedance		25			25		kΩ	
Response Time **		1			1		μsec.	
Reset								
Reset Threshold		1.4	2.0		1.4	2.0	V	
Reset Current		8			10		μA	V_{TR} = 0, V_{RS} = 2V
Impedance		25			25		kΩ	
Response Time **		0.8			0.8		μsec.	
COUNTER SECTION								See Figure 4, V^+ = 5V
Max. Toggle Rate	0.8	1.5			1.5		MHz	V_{RS} = 0, V_{TR} = 5V Measured at Pin 14
Input:								
Impedance		20			20		kΩ	
Threshold	1.0	1.4		1.0	1.4		V	
Output:								Measured at Pins 1 thru 8
Rise Time		180			180		nsec.	R_L = 3k, C_L = 10 pF
Fall Time		180			180		nsec.	
Sink Current	3	5	8	2	4		mA	V_{OL} ≤ 0.4V
Leakage Current		0.01			0.01	15	μA	V_{OH} = 15V

*Timing error solely introduced by XR-2240, measured as % of ideal time-base period of T = 1.00 RC.
**Propagation delay from application of trigger (or reset) input to corresponding state change in counter output at pin 1.

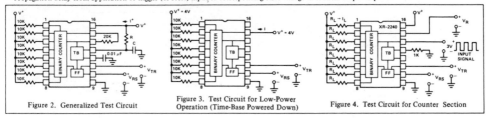

Figure 2. Generalized Test Circuit

Figure 3. Test Circuit for Low-Power Operation (Time-Base Powered Down)

Figure 4. Test Circuit for Counter Section

PRINCIPLE OF OPERATION

The timing cycle for the XR-2240 is initiated by applying a positive-going trigger pulse to pin 11. The trigger input actuates the time-base oscillator, enables the counter section, and sets all the counter outputs to "low" state. The time-base oscillator generates timing pulses with its period, T, equal to 1 RC. These clock pulses are counted by the binary counter section. The timing cycle is completed when a positive-going reset pulse is applied to pin 10.

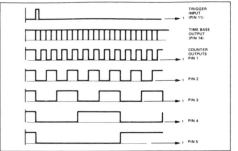

Figure 5. Timing Diagram of Output Waveforms

Figure 5 gives the timing sequence of output waveforms at various circuit terminals, subsequent to a trigger input. When the circuit is at reset state, both the time-base and the counter sections are disabled and all the counter outputs are at "high" state.

In most timing applications, one or more of the counter outputs are connected back to the reset terminal, as shown in Figure 6, with S_1 closed. In this manner, the circuit will start timing when a trigger is applied and will automatically reset itself to complete the timing cycle when a programmed count is completed. If none of the counter outputs are connected back to the reset terminal (switch S_1 open), the circuit would operate in its astable or free-running mode, subsequent to a trigger input.

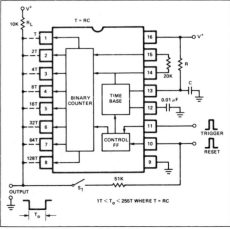

Figure 6. Generalized Circuit Connection for Timing Applications (Switch S_1 Open for Astable Operations, Closed for Monostable Operations)

PROGRAMMING CAPABILITY

The binary counter outputs (pins 1 through 8) are open-collector type stages and can be shorted together to a common pull-up resistor to form a "wired-or" connection. The combined output will be "low" as long as any one of the outputs is low. In this manner, the time delays associated with each counter output can be *summed* by simply shorting them together to a common output bus as shown in Figure 6. For example, if only pin 6 is connected to the output and the rest left open, the total duration of the timing cycle, T_O, would be 32T. Similarly, if pins 1, 5, and 6 were shorted to the output bus, the total time delay would be $T_O = (1+16+32)\,T = 49T$. In this manner, by proper choice of counter terminals connected to the output bus, one can program the timing cycle to be: $1T \leq T_O \leq 255T$, where $T = RC$.

TRIGGER AND RESET CONDITIONS

When power is applied to the XR-2240 with no trigger or reset inputs, the circuit reverts to "reset" state. Once triggered, the circuit is immune to additional trigger inputs, until the timing cycle is completed or a reset input is applied. If both the reset and the trigger controls are activated simultaneously, trigger overrides reset.

DESCRIPTION OF CIRCUIT CONTROLS

COUNTER OUTPUTS (PINS 1 THROUGH 8)

The binary counter outputs are buffered "open-collector" type stages, as shown in Figure 15. Each output is capable of sinking ≈ 5 mA of load current. At reset condition, all the counter outputs are at high or non-conducting state. Subsequent to a trigger input, the outputs change state in accordance with the timing diagram of Figure 5.

The counter outputs can be used individually, or can be connected together in a "wired-or" configuration, as described in the Programming section.

RESET AND TRIGGER INPUTS (PINS 10 AND 11)

The circuit is reset or triggered with positive-going control pulses applied to pins 10 and 11. The threshold level for these controls is approximately two diode drops (≈ 1.4V) above ground.

Minimum pulse widths for reset and trigger inputs are shown in Figure 10. Once triggered, the circuit is immune to additional trigger inputs until the end of the timing cycle.

MODULATION AND SYNC INPUT (PIN 12)

The period T of the time-base oscillator can be modulated by applying a dc voltage to this terminal (see Figure 13). The time-base oscillator can be synchronized to an external clock by applying a sync pulse to pin 12, as shown in Figure 16. Recommended sync pulse widths and amplitudes are also given in the figure.

HARMONIC SYNCHRONIZATION

Time-base can be synchronized with *integer multiples or harmonics* of input sync frequency, by setting the time-base period, T, to be an integer multiple of the sync pulse period, T_S. This can be done by choosing the timing components R and C at pin 13 such that:

$$T = RC = (T_S/m) \text{ where}$$

$$m \text{ is an integer, } 1 \leq m \leq 10.$$

Figure 17 gives the typical pull-in range for harmonic synchronization, for various values of harmonic modulus, m. For $m < 10$, typical pull-in range is greater than $\pm 4\%$ of time-base frequency.

TYPICAL CHARACTERISTICS

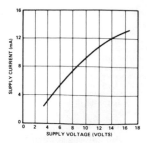

Figure 7. Supply Current vs. Supply Voltage in Reset Condition (Supply Current Under Trigger Condition is ≈0.7 mA less)

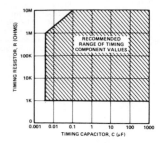

Figure 8. Recommended Range of Timing Component Values

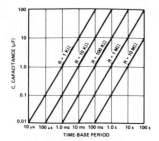

Figure 9. Time-Base Period, T, as a Function of External RC

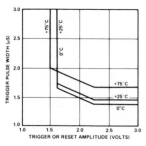

Figure 10. Minimum Trigger and Reset Pulse Widths at Pins 10 and 11

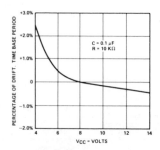

Figure 11. Power Supply Drift

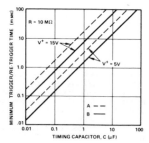

Figure 12.
A) Minimum Trigger Delay Time Subsequent to Application of Power
B) Minimum Re-trigger Time, Subsequent to a Reset Input

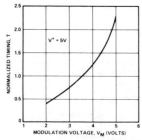

Figure 13. Normalized Change in Time-Base Period As a Function of Modulation Voltage at Pin 12

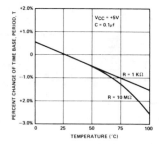

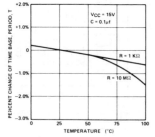

Figure 14. Temperature Drift of Time-Base Period, T

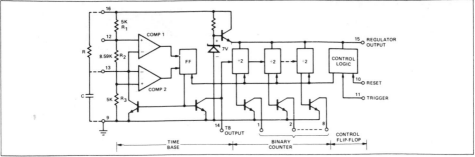

Figure 15. Simplified Circuit Diagram of XR-2240

Figure 16. Operation with External Sync Signal.
(a) Circuit for Sync Input
(b) Recommended Sync Waveform

TIMING TERMINAL (PIN 13)

The time-base period T is determined by the external R-C network connected to this pin. When the time-base is triggered, the waveform at pin 13 is an exponential ramp with a period $T = 1.0$ RC.

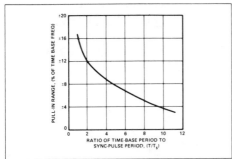

Figure 17. Typical Pull-In Range for Harmonic Synchronization

TIME-BASE OUTPUT (PIN 14)

Time-Base output is an open-collector type stage, as shown in Figure 15 and requires a 20 KΩ pull-up resistor to Pin 15 for proper operation of the circuit. At reset state, the time-base output is at "high" state. Subsequent to triggering, it produces a negative-going pulse train with a period $T = RC$, as shown in the diagram of Figure 5.

Time-base output is internally connected to the binary counter section and also serves as the input for the external clock signal when the circuit is operated with an external time-base.

The counter input triggers on the negative-going edge of the timing or clock pulses applied to pin 14. The trigger threshold for the counter section is $\approx +1.5$ volts. The counter section can be disabled by clamping the voltage level at pin 14 to ground.

Note:
Under certain operating conditions such as high supply voltages ($V^+ > 7V$) and small values of timing capacitor ($C < 0.1 \mu F$) the pulse-width of the time-base output at pin 14 may be too narrow to trigger the counter section. This can be corrected by connecting a 300 pF capacitor from pin 14 to ground.

REGULATOR OUTPUT (PIN 15)

This terminal can serve as a V^+ supply to additional XR-2240 circuits when several timer circuits are cascaded (See Figure 20), to minimize power dissipation. For circuit operation with external clock, pin 15 can be used as the V^+ terminal to power-down the internal time-base and reduce power dissipation.

When the internal time-base is used with $V^+ \leq 4.5V$, pin 15 should be shorted to pin 16.

APPLICATIONS INFORMATION

PRECISION TIMING (Monostable Operation)

In precision timing applications, the XR-2240 is used in its monostable or "self-resetting" mode. The generalized circuit connection for this application is shown in Figure 18.

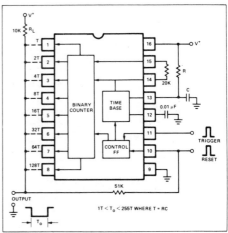

Figure 18. Circuit for Monostable Operation
($T_0 = NRC$ where $1 \leq N \leq 255$)

361

The output is normally "high" and goes to "low" subsequent to a trigger input. It stays low for the time duration T_0 and then returns to the high state. The duration of the timing cycle T_0 is given as:

$$T_0 = NT = NRC$$

where $T = RC$ is the time-base period as set by the choice of timing components at pin 13 (See Figure 9). N is an integer in the range of:

$$1 \leq N \leq 255$$

as determined by the combination of counter outputs (pins 1 through 8) connected to the output bus, as described below.

PROGRAMMING OF COUNTER OUTPUTS: The binary counter outputs (pins 1 through 8) are open-collector type stages and can be shorted together to a common pull-up resistor to form a "wired-or" connection where the combined output will be "low" as long as any one of the outputs is low. In this manner, the time delays associated with each counter output can be summed by simply shorting them together to a common output bus as shown in Figure 18. For example, if only pin 6 is connected to the output and the rest left open, the total duration of the timing cycle, T_0, would be 32T. Similarly, if pins 1, 5, and 6 were shorted to the output bus, the total time delay would be $T_0 = (1+16+32) T = 49T$. In this manner, by proper choice of counter terminals connected to the output bus, one can program the timing cycle to be: $1T \leq T_0 \leq 255T$.

ULTRA-LONG DELAY GENERATION

Two XR-2240 units can be cascaded as shown in Figure 19 to generate extremely long time delays. In this application, the reset and the trigger terminals of both units are tied together and the time base of Unit 2 disabled. In this manner, the output would normally be high when the system is at reset. Upon application of a trigger input, the output would go to a low state and stays that way for a total of $(256)^2$ or 65,536 cycles of the time-base oscillator.

PROGRAMMING: Total timing cycle of two cascaded units can be programmed from $T_0 = 256RC$ to $T_0 = 65,536RC$ in 256 discrete steps by selectively shorting any one or the combination of the counter outputs from Unit 2 to the output bus.

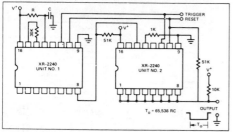

Figure 19. Cascaded Operation for Long Delay Generation

LOW-POWER OPERATION

In cascaded operation, the time-base section of Unit 2 can be powered down to reduce power consumption, by using the circuit connection of Figure 20. In this case, the V^+ terminal (pin 16) of Unit 2 is left open-circuited, and the second unit is powered from the regulator output of Unit 1, by connecting pin 15 of both units.

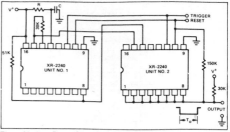

Figure 20. Low-Power Operation of Cascaded Timers

ASTABLE OPERATION

The XR-2240 can be operated in its astable or free-running mode by disconnecting the reset terminal (pin 10) from the counter outputs. Two typical circuit connections for this mode of operation are shown in Figure 21. In the circuit connection of Figure 21(a), the circuit operates in its free-running mode, with external trigger and reset signals. It will start counting and timing subsequent to a trigger input until an external reset pulse is applied. Upon application of a positive-going reset signal to pin 10, the circuit reverts back to its rest state. The circuit of Figure 21(a) is essentially the same as that of Figure 6, with the feedback switch S_1 open.

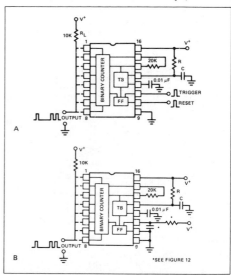

Figure 21. Circuit Connections for Astable Operation
(a) Operation with External Trigger and Reset Controls
(b) Free-running or Continuous Operation

The circuit of Figure 21(b) is designed for continuous operation. The circuit self-triggers automatically when the power supply is turned on, and continues to operate in its free-running mode indefinitely.

In astable or free-running operation, each of the counter outputs can be used individually as synchronized oscillators; or they can be interconnected to generate complex pulse patterns.

BINARY PATTERN GENERATION

In astable operation, as shown in Figure 21, the output of the XR-2240 appears as a complex pulse pattern. The waveform of the output pulse train can be determined directly from the timing diagram of Figure 5 which shows the phase relations between the counter outputs. Figure 22 shows some of these complex pulse patterns. The pulse pattern repeats itself at a rate equal to the period of the *highest* counter bit connected to the common output bus. The minimum pulse width contained in the pulse train is determined by the *lowest* counter bit connected to the output.

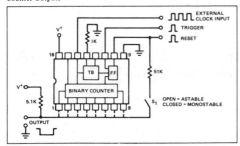

Figure 22. Binary Pulse Patterns Obtained by Shorting Various Counter Outputs

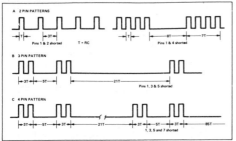

Figure 23. Operation with External Clock

OPERATION WITH EXTERNAL CLOCK

The XR-2240 can be operated with an external clock or time-base, by disabling the internal time-base oscillator and applying the external clock input to pin 14. The recommended circuit connection for this application is shown in Figure 23. The internal time-base can be de-activated by connecting a 1 KΩ resistor from pin 13 to ground. The counters are triggered on the negative-going edges of the external clock pulse. For proper operation, a minimum clock pulse amplitude of 3 volts is required. Minimum external clock pulse width must be ≥ 1 μS.

For operation with supply voltages of 6V or less, the internal time-base section can be powered down by open-circuiting pin 16 and connecting pin 15 to V$^+$. In this configuration, the internal time-base does not draw any current, and the over-all current drain is reduced by ≈ 3 mA.

FREQUENCY SYNTHESIZER

The programmable counter section of XR-2240 can be used to generate 255 discrete frequencies from a given time base setting using the circuit connection of Figure 24. The output of the circuit is a positive pulse train with a pulse width equal to T, and a period equal to (N+1) T where N is the programmed count in the counter.

The modulus N is the *total count* corresponding to the counter outputs connected to the output bus. Thus, for example, if pins 1, 3 and 4 are connected together to the output bus, the total count is: N=1+4+8=13; and the period of the output waveform is equal to (N+1) T or 14T. In this manner, 256 different frequencies can be synthesized from a given time-base setting.

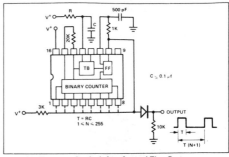

Figure 24. Frequency Synthesis from Internal Time-Base

SYNTHESIS WITH HARMONIC LOCKING: The harmonic synchronization property of the XR-2240 time-base can be used to generate a wide number of discrete frequencies from a given input reference frequency. The circuit connection for this application is shown in Figure 25. (See Figures 16 and 17 for external sync waveform and harmonic capture range.) If the the time base is synchronized to (m)th harmonic of input frequency where $1 \leq m \leq 10$, as described in the section on "Harmonic Synchronization", the frequency f_O of the output waveform in Figure 25 is related to the input reference frequency f_R as:

$$f_o = f_R \frac{m}{(N+1)}$$

where m is the harmonic number, and N is the programmed counter modulus. For a range of $1 \leq N \leq 255$, the circuit of Figure 25 can produce 2550 different frequencies from a single fixed reference.

One particular application of the circuit of Figure 25 is generating frequencies which are not harmonically related to a reference input. For example, by choosing the external R-C to set m = 10 and setting N = 5, one can obtain a 100 Hz output frequency synchronized to 60 Hz power line frequency.

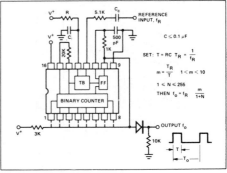

Figure 25. Frequency Synthesis by Harmonic Locking to an External Reference

STAIRCASE GENERATOR

The XR-2240 Timer/Counter can be interconnected with an external operational amplifier and a precision resistor ladder to form a staircase generator, as shown in Figure 26. Under reset condition, the output is low. When a trigger is applied, the op. amp. output goes to a high state and generates a negative going staircase of 256 equal steps. The time duration of each step is equal to the time-base period T. The staircase can be stopped at any desired level by applying a "disable" signal to pin 14, through a steering diode, as shown in Figure 26. The count is stopped when pin 14 is clamped at a voltage level less than 1.4V.

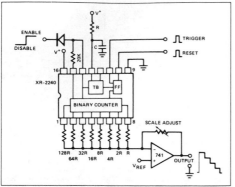

Figure 26. Staircase Generator

DIGITAL SAMPLE/HOLD

Figure 27 shows a digital sample and hold circuit using the XR-2240. The principle of operation of the circuit is similar to the staircase generator described in the previous section. When a "strobe" input is applied, the RC low-pass network between the reset and the trigger inputs of XR-2240 causes the timer to be first reset and then triggered by the same strobe input. This strobe input also sets the output of the bistable latch to a high state and activates the counter.

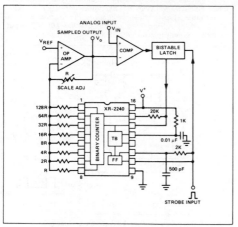

Figure 27. Digital Sample and Hold Circuit

The circuit generates a staircase voltage at the output of the op. amp. When the level of the staircase reaches that of the analog input to be sampled, comparator changes state, activates the bistable latch and stops the count. At this point, the voltage level at the op. amp. output corresponds to the sampled analog input. Once the input is sampled, it will be held until the next strobe signal. Minimum re-cycle time of the system is ≈ 6 msec.

ANALOG-TO-DIGITAL CONVERTER

Figure 28 shows a simple 8-bit A/D converter system using the XR-2240. The operation of the circuit is very similar to that described in connection with the digital sample/hold system of Figure 15. In the case of A/D conversion, the digital output is obtained in parallel format from the binary counter outputs, with the output at pin 8 corresponding to the most significant bit (MSB). The re-cycle time of the A/D converter is ≈ 6 msec.

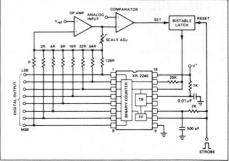

Figure 28. Analog-To-Digital Converter

ORDER INFORMATION

Part Number	Operating Temperature Range	Timing Error	Package
XR-2240M	$-55°C$ to $+125°C$	2% max	Ceramic
XR-2240N	$0°C$ to $+75°C$	2% max	Ceramic
XR-2240P	$0°C$ to $+75°C$	2% max	Plastic
XR-2240CN	$0°C$ to $+75°C$	5% max	Ceramic
XR-2240CP	$0°C$ to $+75°C$	5% max	Plastic

index